DOES THIS PLUG INTO THAT?

Other Books by Eric Taub

Gaffers, Grips, and Best Boys: Who Does What in the Making of a Motion Picture

Taurus: The Making of the Car That Saved Ford

Contributor to:

The New York Times Circuits: How Electronic Things Work

The New York Times Practical Guide to Practically Everything: The Essential Companion to Everyday Life

DOES THIS PLUG INTO THAT?

SIMPLIFY YOUR ELECTRONIC LIFE

Eric Taub

Andrews McMeel Publishing

Kansas City • Sydney • London

Andrews McMeel Publishing, LLC
an Andrews McMeel Universal company
1130 Walnut Street, Kansas City, Missouri 64106

www.andrewsmcmeel.com

14 15 16 17 18 MLY 10 9 8 7 6 5 4 3 2

ISBN: 978-1-4494-2183-0

Library of Congress Control Number: 2012954169

To Carol, the love of my life

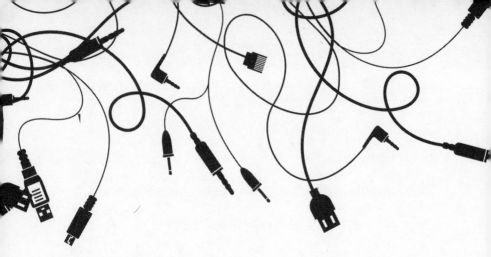

"Technology makes it possible for people to gain control over everything, except over technology."

—John Tudor

"Once a new technology rolls over you, if you're not part of the steamroller, you're part of the road."

—Stewart Brand

CONTENTS

Contents

Contents

Contents

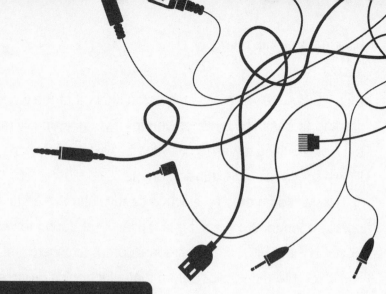

Introduction

Does this plug into that? How many times have you asked yourself that question?

Probably often. If technology is supposed to change our lives for the better, then why is so much of it such a pain to operate?

Perhaps with the exception of Apple, most consumer electronics companies seem to have no idea how the nontechie world lives. Just because they get it, they think everyone else should as well. And if you don't, well, who cares about you anyway?

Kids get it. Children not even able to talk know how to swipe their fingers across cell phones, imitating their parents looking at pictures on their Android smartphones or iPhones. But if you're not of that generation, chances are it's all a bit overwhelming, and if you've never heard of an Android phone, then it's definitely overwhelming. You know that the world is starting to pass you by when you tell a younger family member how exciting it was to get your first extension phone, and he asks, "What's an extension phone?"

The world entered an electronics revolution when technology changed from analog to digital. That began to widely occur in the 1980s, when personal computers were first introduced to the public. Before that, electronics progressed slowly. You may have graduated from a standard phonograph, to a hi-fi record player, to a stereo system, but even as you did, how they operated remained the same. Reel-to-reel audiotape, cassette players, and snapshot cameras were easy to use and easy to understand.

The switch to digital—the encoding of all information as a series of ones and zeros—brought with it a level of previously unimaginable features and complexity. Not only can the sounds and visuals of life be recorded digitally, but now digital attributes can course throughout an entire product.

For example, using an Android phone, you can take a high-resolution photograph and then, while viewing it, alter the tones and apply various effects; you can make the photo look old, saturate the colors, or turn it into a cartoon-like image. You can do these things because you can manipulate a digital copy of the image while viewing it on a screen.

Similarly, you can cut and paste text on screen, change the font, size, and color, and add a beautiful, visually complex border to a manuscript you're typing on a computer. Using a typewriter, the best you could do was to physically cut out text and then move it by taping it to a new location on another physical piece of paper.

Although those examples may be obvious today, they illustrate the power of the digital world; by turning everything into digital code, not only can you create works of art in a device the size of a pack of cigarettes, but you can turn on your oven or program your HDTV recorder from across the globe. You can recreate the sound of a large concert hall in your apartment. And one day, you'll be able to take the activities you've started at home—whether that's a movie you're watching, a book you're reading, or your blood pressure—and continue them as you move throughout your day, from the house to the car, to a flight across the country.

With each added feature comes an exponential increase in complexity, which means that learning a new product often becomes a more involved and convoluted task. But if you're the type of person who wants to stay on top of all the latest gadgets and trends and technologies, there's an easy way to do it.

Just read this book.

Does This Plug into That? isn't an idiot's guide to technology, because if you're reading this, you're probably not an idiot. Rather, you're probably someone who wants to use a lot of technology that's popular today, but you want it to just work, without having to know a lot of acronyms, arcane terms, and the theory behind the technology.

Think about it: You know how to drive a car, but you probably don't know how to fix one. When TVs were picture tube sets, you knew how to change channels, even if you didn't know how the picture got on the screen. So why are you now expected to know which cable to use to connect a Blu-ray player to your TV or understand incomprehensible instructions to set up a wireless home network?

When I was researching my book *Taurus: The Making of the Car That Saved Ford*, the company's head of design asked me whether I knew how they decided where on the dashboard to put the air conditioner vent. I didn't.

"We don't decide," the designer told me. "The engineers decide to put it wherever it's easiest for them to place it." The company was sacrificing the ability to create a good design, one that would please a customer, in order to make life a little easier for its engineers. That's too often the state of today's technology pleasures. The engineers design the products and write the manuals, and then they don't understand why no one can figure out how to use them.

If you want an understanding of how things work and what obscure abbreviations like BD-Live, HDMI, and TCP/IP actually mean, don't read this book. But if you want a helping hand to simply and easily get things up and running, a technological *Guide for the Perplexed* to easily integrate gadgets into your life, and an understanding of why in the end any of this is important, you've come to the right place.[1]

1

The Computer: It's What Makes Everything Possible

> "I was the first person at the USC School of Religion to do my Ph.D. qualifying exams on a computer. People thought this was cutting edge.
>
> Technology for me is simply a cost/benefit analysis. I love the Internet. It's right next to Prometheus[2] in importance. I get to meet other peoples' minds. In two hours I can become fairly well educated. It's electrifying."
>
> —*Rabbi Mordecai Finley, Los Angeles*

> "A computer is like an Old Testament god, with a lot of rules and no mercy."
>
> —*Joseph Campbell*

The home computer has long since passed from an expensive luxury to a necessary tool for everyone in modern society. Whether you live in the United States or Uzbekistan, not having a computer cuts you out of the mainstream of life.

Young adults in Vietnam create animated films in their homes for Hollywood studios and e-mail the files back to North America. Students in undeveloped countries are using inexpensive laptops powered by solar cells to help them learn. And of course, everyone is exposed to this through Facebook, Twitter, and many other social media Web sites.

Not owning a computer, or some sort of computing device such as a smartphone or tablet, is as debilitating as not owning a telephone eventually became in the twentieth century. You simply can't communicate effectively without one.

The debate over what type of computer to buy has been raging ever since the dawn of home computers. Screen size, included software, and, with the debut of Apple's Mac in 1984, the type of operating system continue to be matters of heated discussion. Both camps, Apple aficionados and PC lovers, have good points about why their technology is better. Which one you choose depends on which solid reasons to buy apply to your situation.

For example, if the only thing you like about one particular laptop is that it has a longer battery life but you never use a laptop for more than a few minutes at a time, then battery life shouldn't be important to you. That applies to all electronic products: Don't allow yourself to be wowed by features you'll never or rarely use. Look at a product's specs, test its performance, and buy based on *your* needs, not those of the salesperson or manufacturer.

Why Shouldn't I Buy a Windows Computer? It Seems As If Everyone Uses One

Nearly everyone does. If you like being one of the crowd, you'll certainly feel more comfortable using a Windows machine—or an Apple iPad or iPhone. Being part of the majority has its advantages. You know that there will be plenty of people to whom you can turn when you run into a problem. The more popular a technology is, the more people will work on it, understand it, and, in the case of computers, write software programs for it.

In the days before digital video recorders, when videotape recorders were in their ascendancy, there was an early battle between videotape formats. The system that eventually became the standard, known as VHS, was actually technically inferior in terms of picture quality to its rival, called Beta. But VHS won the battle for one simple reason: You could record up to two hours on a VHS tape, giving consumers the ability to capture an entire feature film on one VHS cassette but not on a Beta.

It didn't matter that most people actually *didn't* record feature films; just knowing that one could do it was comforting to enough people that VHS, with its poorer picture quality, eventually displaced Beta and became the single standard for home videotape recording. For several years, you could still buy a Beta machine and a dwindling supply of prerecorded Beta tapes. But enough consumers decided that wasn't the smart bet, because there was safety in numbers if one chose VHS.

In the same way, there is definitely safety in numbers if you choose a computer that uses the Windows operating system. And for many people that is very important.

Is One Type of Computer Better to Use on the Internet?

It doesn't matter what type of computer you use. Macs and Windows machines are equally good for searching the Internet; downloading files, music, and movies; watching TV shows; and anything else you want to do. Both hardware platforms offer a range of Internet browsers, the programs that make using the Internet possible. Macs come with Apple's own browser, Safari, but you can also use a range of others, including Google's Chrome, Firefox, or Opera, to name a few. Windows machines can also use their own versions of Chrome, Firefox, and Opera, and they come preinstalled with Microsoft's Internet Explorer.

Which Is the Best Internet Browser?

Each browser has its own fans. They all display Web sites in pretty much the same manner. The difference is how quickly each page is drawn and what ancillary features they offer. For example, with Apple's Safari you can save pages in a sidebar for reading later, even when you're not online (although these days, most people are continuously connected). Firefox gives you easy access to quick searches, but Google says its browser launches faster than the competition's.

Which browser you choose is a matter of personal preference. In the end, they all work about equally well.

The Windows Operating System Has More Programs Available. More Choice Is Better, Right?

Studies have shown that more choice doesn't bring more customer satisfaction; it actually brings more anxiety. Too many choices lead to confusion, and too few obviously create a feeling of scarcity. I once entered a small food shop in a village in Belarus in the former Soviet Union and found just one canned vegetable: a brand of tinned peas from Bulgaria, stacked in a pyramid on an otherwise empty shelf to take up more room. I was not happy with my options.

Apple's iPhone store offers around one million apps.[3] But in 2012, just the top 100 generated one third of all the money![4]

You won't use many of the hundreds of thousands of software applications that exist; in fact, you probably won't use more than three or four. Even the app developers complain that with so many apps, it's very hard for theirs to be discovered.

It's the same with computer software. You'll find many more software titles for the Windows PC, but you'll only use a handful. You'll probably want a Web browser (e.g., Internet Explorer, Firefox, Google Chrome, or Apple's Safari), an office suite (e.g., Microsoft Office), and easy-to-use music and video storage programs, such as Apple's iTunes and iPhoto.

Aren't Apple Computers Much More Expensive Than Windows Machines?

Windows machines can be made by anyone, so the competition between the manufacturers drives down prices. You can get a plain Jane computer running Windows for a lot less than a Mac.

I Hear That Apple Computers Are for Artistic People, and I Consider Myself to Be Creative. Will I Be Stymied by a Windows Computer?

Probably not. The little-known secret is that for almost every computer use, there's either a version of the same software that works on both Apple and Windows machines, or there's an equivalent program from the same or another developer that can do an equally good job.

For example, Adobe makes Photoshop in both Mac and Windows versions; Microsoft's famous Office suite of programs also exists for both operating systems (although the designs are somewhat different). iTunes comes in Mac and Windows versions. Although Apple's iPhoto is only for Macs, there are Windows programs (such as Google's free Picasa) that do the same job of storing photos and allowing users to manipulate them.

So Why Shouldn't I Buy a Windows-Compatible PC?

One negative feature of Windows computers is that they attract many more viruses than Macs. Computer hackers around the world are always probing the Internet, looking for vulnerable computers they can infect with invisible programs that can search your computer for hidden account names and passwords, surreptitiously harness your computer to serve up harmful data to other unsuspecting users, or simply inflict wanton mayhem on your machine just for fun.

There are many more viruses that affect Windows computers than Macs. Whether that's so because Windows machines are an easier target or whether it's because Windows machines are more vulnerable is beside the point.

The bottom line: If you use a Windows machine, be prepared to find your computer under attack from evildoers, bad people around the world who will try to infect your computer with programs to steal your personal information, freeze your machine just because they can, and even recruit your machine in an unwillingly conscripted army of computers to attack Web sites and bring them down.

But I'll Save a Lot of Money with a Windows Machine, Right?

Given that anyone can manufacture a PC designed to run Windows, everyone has, and the results are not always good. Just because you got a PC on the cheap doesn't mean you got a bargain.

Aren't Windows Machines Much Easier to Use Than They Used to Be?

Yes, they are. The problem is that although Windows has gotten much easier to use now that it performs more like a Mac, many of its commands remain nonintuitive. If you have trouble remembering routines, a Windows machine may not be for you.

What Are the Reasons to Buy a Mac?

Macs are easier to use because they're easier to figure out. Many applications, including Apple's iPhoto photo storage program and iTunes (also available for Windows PCs), can be learned without reading any manuals. Apple's Time Machine backup system automatically backs up your hard drive, keeping multiple copies of its contents for months at a time so you can easily find a copy of a file that you created a year before.

Aren't All Computers Built by the Same Handful of Chinese Companies?

They may be, but each company that sells Windows PCs has its own set of build quality standards and tolerances it will accept. When you buy a Mac, you buy a computer made by Apple (or at least by Apple's contracted manufacturers),

The Computer: It's What Makes Everything Possible

because only Apple can make computers that run its operating system. So you don't need to worry much about shabby build quality.

Is It Really True That Macs Don't Get Viruses? That Sounds Like an Urban Myth.

Unlike Windows PCs, Macs can be infected by only a few viruses. In twenty-seven years of using Macs, I've had two viruses, neither of which did any apparent damage to my computer, and they were easy to remove with a free antivirus program.

For years, the tech industry has been predicting that Macs would eventually get as many viruses as Windows machines as soon as the population of Mac users increased. The thought was that the cost–benefit ratio for the hacking community would improve and make it profitable to write more viruses.

In 2010, Apple claimed that 20 percent of all new computers sold ran on the Mac operating system, and other stats say that more than 7 percent of American computer users have Macs, indicating a definite increase from about 2 percent a few years ago.[5] Still, only a handful of viruses have attacked Macs, and most affect only a small minority of users.

Can I Get the Programs I Need for a Mac?

As stated earlier, you may not find every Windows program in a Mac version, but most are available. Those that aren't typically have a similar Mac program that does much the same.

Computers Scare Me. What If the Computer Crashes?

It's been a long time since computers passed from interesting but superfluous tools to must-have devices. That's true whether you live in Seattle or Soweto. Today, everything happens on the Internet, from education to paying bills to watching TV. But computers do crash. When Apple first introduced its OS X operating system, the company claimed it was so advanced that even if a program did crash, it would not freeze the entire machine—only that program. It wasn't true. It hasn't happened often, but there have been times when one program freezing did cause my entire machine to lock up.

The Most Important Thing You Can Do to Fix Your Computer, Smartphone, iPad, etc.

When your computer stops working—whether it freezes, a program crashes, or it doesn't respond in some other normal way—there's one thing that usually works to set things straight: Restart it.

Restarting your PC (and your smartphone, tablet, portable navigation system, and most other digital devices) can cure many ills. It's like putting digital Drano in your machine: It flushes out the gunk and restores functionality.

A real-life testimonial: I was unable to get my iPad to send mail, even though all the settings were correct. I deleted my account and reentered all the settings properly multiple times. The result—nothing.

Then I deleted my account once more, shut off the iPad, and reentered all the information after restarting. Everything was back to normal.

What If Restarting My Computer Doesn't Solve the Problem?

There are too many ways a computer can go wrong to list them here. However, there is one trick that typically works: Search the Internet for a solution to your particular problem. More often than not, someone else has experienced the same thing. Before throwing up your hands in frustration, simply Google the keywords explaining what your problem is. You'll almost always find many others with the same problem and, usually, a solution. As a last resort, you can hire a live expert. But I've always found the answer I need by searching on the Web.

If I Use My Computer to Store All My Documents, Photos, Movies, and Music, How Do I Find Everything?

With giant hard drives now the norm, it's easy to file away tens of thousands of files and never run out of room. (I've got 40,000 received e-mails on my PC, plus thousands of photographs and scores of programs, among lots of other stuff.) Both Apple and Microsoft have made it pretty simple to locate your files, and their search procedures are similar.

How to Find Things on a Mac

Apple's Spotlight file-finding tool lets you search your entire Mac, or specific folders, for the file you want. You can look for a particular word in a file's title or a word in the body of the file itself.

To use Spotlight, you either click on the picture of the little magnifying glass in the upper right-hand corner of the screen or go to the word "Find" in the File menu, when you're actually in the Finder part of the Mac. (To get to the Finder, you click on the picture of the Mac happy face, which is the leftmost icon in the computer's program dock.)

When you click on the Spotlight magnifying glass, you can type the name of a file or a word that appears in a file. You'll then see a list of appropriate files organized by type: all the matching documents, folders, pictures, and so on. If you click on the first phrase in the list, "Show All in Finder," you'll see your results in a standard list window, with text that indicates where that file (or folder) resides.

You can also use Spotlight to do simple math calculations rather than opening up the calculator application or looking for your battery-operated one that's hidden under a pile of papers on your desk. Enter the calculation in the Spotlight bar using standard mathematical characters (e.g., 2+2, 2x2, 2–2, or 2/2). The first listing under the Spotlight window will be the result of the calculation.

How to Find Items on a PC Running Windows 7 & 8

Point to the upper-right corner of the screen, move the mouse pointer down, and then click "Search." Type the search term in the box. As you type, files containing that word will appear.

If you're looking for a certain app, or setting, or file, then click on that same word (app, setting, file).

In Windows 7, you can also search in specific libraries, Microsoft's term for a collection of file types. There are four libraries shown on the left of the window: documents, music, pictures, and videos. These libraries don't actually contain the files. Rather, each acts as an organizing tool, showing you a list of all the documents, all the music files, and so on. If what you want is definitely music, you can open the music library and type in the search box the specific word that represents the file you're searching for. You can also modify your searches by adding qualifier filters, such as the date the file or folder was created. You can specify a date or simply choose "A long time ago." You can search by exact name; you can exclude a word or search files of only a certain size. All these options are presented to you in the search box; which options you see are determined by the type of file you are looking for.

How Do I Find What I'm Looking For on the Internet?

When searching on the Internet, I operate under this theory: Anything I want to know has already been asked by someone else. Without exaggeration, that has proven to be true 100 percent of the time; that's a good thing, because it makes searching the Internet much easier.

When searching for a topic, whether it's news, an image, or a concept, use keywords without making the search too broad or too specific. For instance, if you are looking for a news report on a Texas judge who expects the United States to become a ward of the United Nations, enter terms such as "Texas," "United Nations," and "takeover." Interested in a particular modestly priced digital camera? Don't just search for "digital cameras"; you'll get too many responses. Instead, search for "mid-range digital camera best reviews."

You can include or exclude certain words to make your searches more accurate. For example, if you want to search for information on the jaguar animal, you can type (without the quote marks) "Jaguar –automobile"—Jaguar minus automobile—to make sure your search results don't include the car by the same name. On the other hand, if you only wanted information on the automobile, you could type "Jaguar + automobile" (again without the quote marks). If you can't remember the various modifying marks, Google has a search page where it will put in the marks for you (google.com/advanced_search).

Although Google has become the search engine of choice for most, also try Bing (bing.com) and Yahoo! (yahoo.com) to expand your results.

How Can I Make Sure I Don't Lose All My Digital Stuff? I Keep Hearing Horror Stories About Getting Hacked. Or What If My Computer Is in a Fire?

There are three—not two—things that are guaranteed in life: death, taxes, and the loss of the information stored on your computer or portable device.

It hasn't happened to you yet? Neither has death. But both will.

Hard drives fail (often unexpectedly). Lightning storms fry computer components. Portable devices such as smartphones just up and die. To prevent the disaster that could happen when you lose all your family photographs, your work documents, and your favorite music that you've spent hours transferring to your computer, there are two things you should do:

- Back up everything to an external hard drive.
- Back up your essential files to the "cloud."

If you're an Apple user, it's easy (perhaps you've heard that refrain before). Macs come with a program called Time Machine. Plug an external hard drive (using the included USB cable that comes with a hard drive) into your Mac's USB port. Then you just go to Time Machine in the System Preferences application (to get to it, choose "System Preferences" under the [Apple] menu in the upper left of the screen, slide the on/off bar to "on," go to "select disk," and select your external hard drive, and you're done).

Windows 7 has a similar feature. In the Windows 7 operating system, click the Start button. Then open Control Panel; once there, click on "System and Maintenance," and then click on "Backup and Restore." If this is your first time, click on "Set Up Backup" and follow the instructions. Windows 8 calls its automated file backup system "File History." To activate it, use your mouse to point to the upper-right corner of the screen, pull the mouse pointer down to show the Search box, and type in "File History." Then click to turn it on.

What Happens If My Backup Hard Drive Breaks?

Backing up at home is great, but what happens if your home is flattened by an earthquake or a tornado? Or your home is flooded and your hard drive gets soaked?

The answer is in the cloud.

What's the Cloud?

You may have heard the terms *cloud computing, cloud-based music,* and *cloud-based backup.* What they mean is simple: Files are stored in a bank of computers in some remote location. Typically, companies that provide cloud services have "server farms," scattered all over the world. These farms don't have cows or sheep, just hundreds or thousands of hard drives electronically tied together that store information. When you read your e-mail from Google or Yahoo! or Apple's iCloud service, those messages are stored in the cloud, at the companies' server farms.[6]

Some companies are even storing their computer programs in the cloud. So rather than having to buy and install a program on your PC, you access a master copy that's on a remote server somewhere, via your Internet connection. That way, you're always using the latest version of the program (no updating necessary), and you don't have to worry that the program might get electronically corrupted or that you might lose the master disk with which you installed it.

Apple's iCloud service not only holds your e-mail (if you use an "@me.com" or an "@icloud.com" address), but it can also be used to store the digital files you've purchased from the Apple iTunes store. The servers even hold copies of music that you may have transferred from CDs to your computer; if you lose your copies you can download new copies from Apple (for an annual fee) or access the same music on your smartphone when you're in a distant location.

Cloud servers also make for great alternative backup locations. With a secondary cloud backup, you'll know that your data will always be safe if the backed-up files on your hard drive become inaccessible.

Do You Back Up?

If you're like most people, it's likely that you don't back up your files, either on a local hard drive or in a storage space in the cloud.

According to a study of Western Europeans conducted by Acronis, a backup company, 83 percent of respondents understand the need to back up, but only 15 percent back up once a week or more often, and 29 percent never back up.

Should you back up? Absolutely. Should you back up in multiple locations? As my mother used to say, "It can't hurt." Adding a cloud backup service to your backup is simple; several companies offer similar pricing plans and virtually foolproof instructions. Backing up to a remote location has an added benefit: When a new type of storage format makes hard drives obsolete, as it most definitely will, you won't have to worry about transferring your data to the newest technology.[7]

How Do I Get My Stuff into the Cloud?

It's easy to back up your files to the cloud, and here are two services to consider: Carbonite (carbonite.com) and Mozy (mozy.com). There are others, but these two are among the easiest to use.

For about $60 per year, Carbonite lets you back up one computer, no matter how many files or gazillions of gigabytes you have on that PC or Mac. You don't get to back up additional computers for that price, nor can you back up files on an external hard drive that's connected to that computer. For that privilege, you'll need to pay an additional $40 per year.

Mozy charges $6 per month, or $5.40 per month if you pay for two years in advance. That price buys you 50 gigabytes of file backup, enough for most people. The company's Web site points out that 50 gigabytes equals more than six million e-mails, 7,500 photos, 8,500 music files, or 1,000 videos. Of course, that's just a rough estimate. If your movies are all six hours long or your photos were shot at a very high resolution, that number will drop.

The Computer: It's What Makes Everything Possible

Hint: How to Reduce Your Backup Size

Don't back up your applications (programs such as Microsoft Office or iTunes); if you ever have to restore your entire computer because of a catastrophic failure, you can restore your applications from the original CDs, or you can download them again if you purchased them over the Internet. For safety's sake, you can always make backup copies of your programs on recordable CDs or DVDs as soon as you buy them.

With both Carbonite and Mozy, you simply download an application onto your computer. The programs walk you through the setup, and then they back up your computers automatically and continually.

Is the Cloud Just for Backing Up My Files?

Cloud backup services aren't just for catastrophes. Once your files are in the cloud, you can grab one when you're away from your main computer that's being backed up.

Let's say you're at your Aunt Martha's house in Evanston, and you want to give her a copy of that great brownie recipe you downloaded from the *New York Times* Web site.[8] If you have the Carbonite or Mozy app installed on your Android or iPhone smartphone or tablet, you can find the file and download it on the spot. If you have a laptop with you, you can do the same thing just by going to the Carbonite or Mozy Web site, signing in, and accessing your file. You can search for it by name in the search box, or find it by navigating through the folders in which it's stored.

2

Computer Printers

When people first began using computers, they printed everything. The idea of reading something on a screen or correcting a written document without using a paper version was anathema to a lot of users. Although the earliest printers typically cost up to $1,000, they were essential for many people. Today, printers that double as scanners and fax machines cost around $85. Their utility has increased, but the necessity of having one has gone in the opposite direction. That's because screens have become ubiquitous and we are now accustomed to viewing and manipulating our digital work without first making a paper copy.

We e-mail photographs and view merchants' bills on the screen. We store our photographs electronically, and when we want our friends to see them, we post them on Facebook or Instagram, show them on our iPads, or transfer them wirelessly from our computer or smartphone to our large-screen HDTV.

Is a Computer Printer Still a Necessity?

At this point, it's easy to get along without a printer. But they still come in handy sometimes. Printing out directions from the Google Maps Web site is often easier than trying to read the map on your smartphone. You can wave your smartphone at the entrance gate to an event to show your electronic ticket, but it's more reliable to print a paper copy in case your cell phone battery dies just as you get to the front of the line.

What Should I Look For in a Printer?

Because printers are now so inexpensive, you can buy a loaded one for a low price. For less than $100, you can get a machine that prints, scans, and faxes.

Look for a printer that comes with its own e-mail address. The point isn't that you can send your printer love letters; with an embedded e-mail address, you can e-mail the printer a document (or photo or whatever), and you can print it directly, without having to first open the document on your computer and then tell the printer to make a copy. With an e-mail address attached to a printer, you can e-mail a document to it from anywhere in the world, and it'll print. That's a great way to send photographs of your trip to Machu Picchu to your parents without asking them to fiddle with e-mail attachments and computer commands.

Today's printers also come with wireless capabilities. You no longer need to tether your printer to your computer with a cable. Instead, a wireless printer lets you send files to the printer without a physical connection between the two. That's handy if your printer is far away from your computer, and it allows other computers in your household to send files to the printer, as long as they're all operating on the same wireless network (and computers in one household typically are). If you have a tablet computer, such as an Android tablet or an iPad, you can print a photograph that you received in an e-mail by just sending the file to the printer wirelessly.

What Are the Newest Features of Today's Printers?

Printers now come with their own apps—applications that add functionality by downloading information from the Internet. For example, with a Google Maps app, you can look up directions right on the printer and then print those directions without using a computer to first access the map.

The Fandango app allows you to look up show times and print tickets from the printer. If you're a *60 Minutes* fan, you can download and print transcripts from the show directly to the printer.

The truth is, though, there's really not that much use for apps on printers. This seems to be another case of an industry running scared. Printer executives see how popular apps are on smartphones and computers, so they figure they'd better put them on printers as well. Soon we'll see apps on electric toothbrushes (they're already on pens). But just because you can do something doesn't mean you should.

Why Is Printer Ink So Expensive?

In 2011, I asked Johan de Nysschen, then-president of Audi USA, why the company charged so much for its small A3 sedan. "Because we can," he answered. And that's why printer ink is so expensive: because companies such as Hewlett-Packard can ask a high price.

Printer giant HP charges a pittance for its printers but a lot more for its ink. It's another application of the razor and razor blade approach: Charge a little for the hardware device, but then charge a lot for the products you need to make that hardware work well. If you make those products better than your competitors', consumers will continue to buy your consumable products. Kodak tried to take a different approach. Its printers cost a bit more, but its ink generally costs less than HP's. However, that strategy failed; Kodak is now out of the printer business.

Because most people don't do much printing, the price difference in ink has become negligible for many. In my own work, I print three to four pages per week. A typical HP cartridge lasts me for months.

Which Printer Brand Should I Buy?

Printers made by big names such as Canon or HP work well, and they are simple to set up: Plug the printer into the computer and the power cord into the wall. The computer automatically recognizes the type of printer you have and presents you with the proper print instruction box when needed.

3

I Just Want to Watch TV!

"I've always been a populist. I'm interested in how to use technology as a way to communicate, play, and let people express themselves."

—*Steve Mayer, co-founder, Atari*

Remember when watching TV was easy? Here's how we did it:

1. Turn on TV.
2. Change channel.
3. Watch show.
4. Go to refrigerator.

Things got a bit more complicated with the invention of the remote control (a great boon for nascent couch potatoes). And then when cable came along, we got a lot more channels, but it also became harder to find what we wanted. We had to navigate program guides and figure out how to record things, first on a VHS machine and on a box the cable or satellite company gives us to store programs.

With giant flat-screen TVs, we don't just want to watch TV; we want to recreate the theater experience in the home: huge HDTVs, multispeaker surround sound, extraordinary picture quality from Blu-ray discs, streaming movies from the Internet right into the set. It all makes for a great experience, but it comes at a price: confusion.

Every new box in your home comes with its own remote. If you're like most of us, you've got a coffee table filled with five or six of them: for the TV, the Blu-ray or DVD player, the video recorder, the Xbox or PlayStation, and the home theater receiver. If you're still living in the past, you may even have one for the VHS machine you never use except to watch the tape of the quiz show you were on in 1970.

Now you've got a flat-screen TV that's probably as big as the bathroom in the house you grew up in, and you're using it to watch high-definition TV. Or are you?

I've got an HDTV. So I'm watching high-def TV, right?

Here's how to find out: Is the picture super-sharp and wide

as well? If so, you're watching high definition. If the screen is wide but the picture itself is squarish, you're not. If you haven't yet bought an HDTV, this is what you need to know.

What Is High-Definition TV?

It's a TV that shows images that are much sharper than on a regular TV. What's the big deal? The sharper the image, the easier and more fun it is to watch. If you're an old-timer, remember how crummy the picture looked when you watched a VHS tape? You may have gotten used to it, but it wasn't very nice to look at. With a sharper image, you can sit closer to the screen, which makes the picture appear bigger and more in your face, like being at the movies. When it comes to TV, bigger and sharper is always better because what you're watching feels more immediate. It comes alive.

How Do I Get HDTV?

1. Buy an HDTV set. If you don't have an HDTV, you can't see high definition.

2. Get an HDTV cable or satellite box. Subscribing to HDTV generally costs more than standard-definition TV, but the cable and satellite companies typically will give you the special HDTV decoder box at no charge, often in exchange for a multiyear commitment. If they won't at first, just ask. Threaten to go to a competitor, and they'll usually give in.

3. Tune to an HDTV channel.

Is Everything on TV in High Definition?

No. If the channel you're watching delivers a wide-screen image on your HDTV, it's in high definition. If the picture looks squarish, like on the old picture tube TVs, then you're just watching standard, low-definition TV, but you're using an HDTV to do so.

The Secret Way to Get Lots of Free HDTV

Remember when TV used to be free? You used rabbit ears or a roof antenna to pick up local channels. It still is free, but now it's even better. With the right-sized antenna you can watch local HDTV channels for free, just like in the old days. Thanks to digital technology, the space that those channels take up on the airwaves can be compressed so they take up less room, or "spectrum." And with that extra space, you can see more channels from each broadcaster.

So channel 2, for example, can be squished down in size and be broadcast along with channel 2.2, 2.3, and so on, up to about five additional ones. Lots of broadcasters are doing this. Not only do you see the main channel you're familiar with, but with a digital antenna you'll also be able to pick up those "subchannels," and on them you'll find music programming, arts, and even 24-hour news.

With a few rare exceptions, unless you get your TV channels using a regular antenna, you won't be able to see any of them. That's because most cable and satellite companies won't carry the subchannels. (Hint: They can't make any money carrying them.)

How Much Do These HDTVs Cost?

A lot less than they did when the technology was new. Twelve years ago, an HDTV cost about $10,000 for a fairly small 34-inch screen. Today, $1,300 can buy you a much better-looking image in a set as big as 60 inches.

So should you just sit tight until the price drops further? There's no point in waiting. Cars may change only slightly year to year, but modern consumer technology changes rapidly. Anyway, the prices are so low that they can't drop that much more any time soon. The fact is that almost every big-name TV manufacturer—Panasonic, Philips, Samsung, and Sony—is losing its shirt selling HDTVs; the prices are so low they actually sell them at a loss just to keep their share of the market.

How to Choose a Good HDTV

The first HDTVs used picture tubes to create their images, just like the old regular TVs. Picture tubes make great images, but man, they are *heavy*. Remember how difficult it was to pick up a large picture tube TV? You either got a friend to help, or you did it yourself and then checked into the hospital for your hernia operation. Picture-tube HDTVs were even heavier—more than 100 pounds. Just delivering the set to your house was about as easy as carrying a piano.

Fortunately, the TV industry came up with two different technologies that dramatically lightened things up: plasma and LCD. If you've heard lots of bad things about plasma—the sets don't last long, they use a lot more power—forget about them now. Whatever shortcomings plasma once had are long gone.

A Few Simple Rules About What TV to Buy

If you want the best picture quality, get a plasma TV.

If you watch TV in a really bright room, get an LCD TV.

Plasma TVs look especially dim in stores like Best Buy and Costco because they're lit by bright fluorescents. To make matters worse for plasma, the LCD TV manufacturers have a special "retail" setting to blast out their screens' brightness on the store floor. The brighter the image, the sexier it looks. And because LCD TVs can be made much brighter than plasmas, they always look better. That's great if you're watching TV in a fluorescent-lit room; but outside of a few apartments in Kazakhstan, I haven't seen many fluorescents in American living rooms. In normally lit rooms—like those in your home—the brightness of a plasma set is just fine.

And you know those "LED TVs" you see advertised all the time? There's actually no such thing as an LED TV. It's a marketing gimmick. "LED TVs" are actually LCD TVs that use LED lights to create the TV's light source. It's just a way for manufacturers to create another category, confuse consumers, and hopefully get people to pay a bit more money.

What Do I Look for If I Want an LCD TV? An LED or an LCD?

All the LCD TVs will soon use LED lamps, so you don't even have to worry about that.

What Size TV Should I Get?

Because HDTVs can display sharper pictures, you can sit closer to the screen. Divide your viewing distance by 1.5 to 3 to figure out the diagonal size of the TV you should get. For example, suppose you're sitting 10 feet from the TV. Ten feet = 120 inches. 120/1.5 = 80, and 120/3 = 40. So buy a TV that has a diagonal screen size between 40 and 80 inches.

What Do I Do with All the Picture Settings on the TV's Menu?

Most people will buy a TV, take it home, turn it on, and figure they're done. That's not surprising; when picture tube TVs were the standard, I was amazed that consumers often seemed perfectly happy to watch an image of someone whose skin color made him look as if he came from another planet. Purple features don't seem to bother some people. It's surprising how many set their TVs to the "vivid" setting, which gives the picture an artificial, supersaturated look.

To make things easy for the average consumer, all TVs today come with a variety of presets, tweaks that tune the picture to look best in different lighting situations.

For example, in addition to "standard," one Panasonic TV offers Vivid, Cinema, Game, Custom, and THX settings. Not to be outdone, and to make the choice even more complex, a Sony model lets you choose from Vivid, Standard, Custom, Cinema, Sports, Game-Standard, Game-Original, Graphics, Photo-Vivid, Photo-Standard, Photo-Original, and Photo-Custom. Whew!

Panasonic recommends that you use Vivid in a bright room, Cinema in a darkened room (the picture is less bright and softer, like a movie), and Standard for "normal viewing conditions," whatever that means. Sony's manual says that Standard should be used "for standard pictures;" Cinema is for "film-based content," and Sports "optimizes picture quality for viewing sports." But what if you're watching a sports movie or a live sporting event? Does that demand a Standard, Cinema, or Sports setting?

Then there's Panasonic's special THX mode, which "faithfully reproduces the image quality that the movie makers intended in order to provide the ultimate cinematic experience at home." So what's the difference between THX and Cinema modes? Officially, THX will give you a professional calibration of sharpness, color, brightness, and contrast, whereas Cinema will optimize the image to make movies look best. Cinema is trying to do the same as THX, but it's not the professional calibration.

In reality, you may not be pleased with either calibration. Choose what looks good to you as long as you're trying to create the best *natural*-looking picture, one that reflects the way the world looks. If your goal is to make people look as weird as possible, I can't help you. But if you want to recreate the world on your TV, what follows are some hints on how to make the adjustments yourself.

What Is THX, and How Did It Get Its Name?

We've all heard that cool sound and seen that giant "THX" logo that often precedes movies in the theaters (you can see and hear it at is.gd/XHyHML. But do you know what THX is? It's not a technology; it's a quality assurance system that "guarantees" that a movie will look or sound the way its creator intended.

The THX system has been adopted by companies such as Panasonic to ensure that the settings on a TV create the ultimate picture quality, not too washed out and not too saturated, with lifelike colors.

The name comes from its inventor, audio engineer Tomlinson Holman, and the "X" stands for "crossover." He created it while working at George Lucas's company, Lucasfilm.

Getting the Best Picture from Your HDTV

If the TV that looked great when you were shopping at Costco looks just average and disappointing now that you've brought it home, this section is for you. Most of today's digital TVs—if you have a flat-panel LCD or plasma TV, your set is digital—can present a great-looking picture, especially if you're watching in high definition. Here's how to tune your TV to get that great look.

Turn Down the Lights

According to Mark Schubin, a famous television engineer and Metropolitan Opera technical consultant, turning down the lights is the number one thing to do. If the room is too bright, you can't maximize the range from the whitest white to the blackest black. The wider the range, the more lifelike the picture will be. We've all seen pictures that are washed out and muddy looking. You don't want that in your home.

If you cannot lower your room's lighting, at least never let the sun directly hit the screen. If that happens, you'll be lucky to see any image at all.

Next, Adjust the TV's Brightness Control

The brightness control adjusts the level of the blacks in the picture. (Shouldn't it be called the blackness control?) If the dark parts of a scene don't look black but instead look muddy, then the picture will look poor. While watching a dark scene in a movie, turn the brightness/picture control down until the detailed areas in a dark part of the frame disappear, then turn it back up until you can just make out some detail.

Now, Adjust the Contrast

You've gotten the blacks to look good. Like those laundry detergent commercials, you need to make sure the whites look white and not gray. But not too white: If you crank up the contrast too much, you'll think you're in a whiteout in Aspen. That's not good.

The simplest way to do the adjustment is to use a pattern on a tuning disc or one that is part of a THX-certified DVD, which will include picture-tuning instructions (see page 43). Otherwise, find a bright scene in a movie—a park on a bright day would work—and then adjust the contrast so you can make out blades of grass, shadows, and folds on clothing.

Color: More Is Not Better

If the colors on your TV screen look so rich you think you just smoked a joint, you've gone too far; too much color starts to look silly, not beautiful. Keep the comic book look for comic books. Make sure people's skin tones look pretty much the way they do in real life. In North America, there are two color adjustment knobs: color and hue. They interact, so after tuning one you may need to retune the other.

Write It All Down

Jot down the numbers of the settings, in case your dog or child steps on the remote and goofs them all up. Also, if you watch TV at different times of the day when the light changes dramatically, you'll want to go through the same process for each lighting condition.

But Why Didn't This Work for My Blu-ray Disc Player?

In olden days—five years ago—setting the picture controls on your TV changed them no matter what you were watching. Today, setting the controls while watching cable TV will often have no impact on the picture when you're watching a DVD or playing a video game.

This is a feature, not a mistake. TV manufacturers figure everything you have connected to the TV will create a different-looking picture; being able to adjust each one inde-pendently will give you more control. Whether you want that control is another matter.

About Those Tuning Discs

To do the best job, you may want to buy a DVD or Blu-ray tuning disc. You insert the disc in your Blu-ray or DVD player, and it takes you through various tuning routines. For example, you'll see a pattern of vertical black and gray bars, and you adjust them so that one of the black bars matches the background of the image. Other test patterns help you set your picture's sharpness and color rendition to match the colors that a scene is intended to show.

Some of the tuning discs you can choose from include *Digital Video Essentials: HD Basics,* the *Avia II Guide to Home Theater,* and the deftly named *Spears & Munsil High-Definition Benchmark Blu-ray Disc Edition.*

THX now sells THX tune-up, a $1.99 tuning app for iOS and Android devices that offers similar tuning routines to the various discs.

If you don't want to shell out the big bucks for a tuning disc or smartphone app, you can get the same setup patterns for free on many DVD and Blu-ray movie titles, such as *Avatar, Star Wars: The Complete Saga,* and *Toy Story.* They and others include THX Optimizer, a range of calibration patterns created by THX Ltd., the company whose familiar certification logo precedes many movies. To get a better sense of what's involved here, go to bit.ly/A5zlix for a primer on display calibration, including what you have to do and what it does.

Filters are included with tuning discs; if you're using a THX pattern that's included on a DVD, you can purchase a set of blue filter glasses from THX's Web site (www.thx.com). Depending on the country in which you live, a pair will cost between $4.50 and $6.00.

THX's Web site also has a list of the more than 300 DVD and Blu-ray titles that have the THX Optimizer calibration program built in. To see it, go to bit.ly/X42Sft.

Sharp and Sharper: DVD vs. Blu-ray

Remember how bad VHS tapes looked, even on a regular picture tube TV?

The invention of DVD dramatically improved picture quality and added additional features, such as multiple audio tracks (in case you wanted to watch an Iranian movie in Persian), deleted scenes, and movie trailers. But in the consumer electronics industry, good enough never is.

The problem with using a DVD connected to an HDTV is that the HDTV can show pictures sharper than the DVD can produce. It's like driving a car with Vaseline on the windshield; your eyes may be sharp, but the lens you're looking through isn't. To solve this problem, Philips, Sony, and others created a high-definition DVD standard called Blu-ray. Now, movies on disc can look as sharp as the images the TV is capable of producing.

The greater capacity of the Blu-ray disc gave filmmakers extra room to put on even more extraneous material you'll never look at, such as movie-related games, timelines of the action, and actor bios. If you connect the Blu-ray player to the Internet, some discs can access even more programming by downloading it while you're watching.

Ever rented an old movie and been forced to watch trailers for films that came out years ago? With a Blu-ray disc played in an Internet-connected Blu-ray player, the trailers you see can be downloaded from the Internet while you're watching them, guaranteeing that you'll be presented only with advertisements for the latest movies you have no desire to see.

Like every other technology, Blu-ray players have come down dramatically in price. You can easily pick one up for around $100 or less; standard DVD players are about $40.

3D TVs: Should I Buy One?

For a few years, 3D TVs were all the rage—at least as far as the TV manufacturers were concerned. In 2010, every manufacturer was pushing 3D as the Next Big Thing, because they needed a Next Big Thing. With TV prices continually plummeting, the companies needed something to boost profits.

Unfortunately, it didn't work out the way they had hoped. A few 3D TV channels launched, a few big 3D shows were produced, but to date, 3D has never really taken off.

Soon, you won't even have to make a decision about buying a 3D TV because 3D will be just another feature that comes with almost all TVs, just like surround sound. Until that time, watch some 3D excerpts in a retail store. Do you like what you see? If you don't, then you may not want to spend the extra money for 3D capability.

How Can I Get 3D TV Programming?

Most cable and satellite companies offer one or more 3D channels. You can also buy 3D Blu-ray discs. If your TV connects to the Internet (more on that later), you may be able to watch 3D programming that's automatically downloaded (or *streamed*) into your TV.

Will My Blu-ray Player Work with 3D Discs, and a Few Other Things You Need to Know About 3D TV

You'll need a 3D-compatible Blu-ray player. They cost a bit more than regular Blu-ray players.

Just like 3D in the movies, some 3D TV looks good, and some doesn't. Certain kinds of 3D programming make a visual spectacle such as a rock concert, an opera, or the opening ceremony of the Olympics look even more spectacular. And some 3D programming makes a bad program look even worse. 3D may give you a wonderful sense of depth and presence, or it may look like a school kid's diorama, just a series of planes at different points in space.

3D is in its infancy, and some of the people creating it haven't figured out how to do it well.

Do I Have to Wear Glasses When I Watch 3D TV?

Yes, but you may not find it such a big deal. After all, you wear glasses when you see a 3D movie in the theater. On the other hand, there is one problem with 3D TV glasses: If you lie down or tilt your head, you'll lose the 3D effect. Also, wearing glasses creates an even more isolating environment when you are watching TV. With glasses on, you'll be less likely to turn to others in the room to have a chat.

Of course, that might be a good thing.

I Just Want to Watch TV!

What Kind of Glasses Are There?

There are two different types of 3D glasses. One requires
power from a small battery in the frame. The other type has
no batteries, just like the ones you use at the movie theater.
In fact, you can use ones from a movie theater with some
3D TVs; they are called "passive" glasses, and the more
expensive battery-operated ones (which are not compatible
with the movie theater glasses) are called "active."

The active glasses cost more—sometimes a lot more.
Some manufacturers charge $150 for a pair. The battery-free
glasses are so cheap the television companies usually throw
in four or more pairs for free.

Pros and Cons of 3D TVs with Active and Passive Glasses

3D TVs designed for use with active (battery-powered) glasses:

- Glasses are expensive (around $60 and as much as $150 a pair).

- 3D picture is sharper than that of battery-free glasses.

- 3D picture is darker than that of a passive-glasses 3D TV.

●●●●●●●●●●●●●●●●●●●●●●●●●●●●

3D TVs designed for use with passive (battery-free) glasses:

- Glasses are usually free and included in the box.

- Picture is not as sharp as with an active-glasses 3D TV.

- Picture is brighter than with an active-glasses 3D TV.

●●●●●●●●●●●●●●●●●●●●●●●●●●●●

Does the 3D TV I Want Use Active (Expensive) or Passive (Cheap) Glasses?

If your 3D TV comes with four or more pairs of glasses in the box, you're probably buying a passive-glasses 3D TV.

I Don't Like Glasses! Why Can't I Get a 3D TV That Doesn't Require Them?

A 3D TV that doesn't require glasses to view is the Holy Grail, and companies are trying hard to perfect one. The technology to make a glasses-free 3D TV is the same as that used with the plastic scenes that often come in Cracker Jack boxes: Move the little picture left and right and you get a 3D effect.

The problem is that it's difficult to make that effect work on a large screen so that lots of people sitting in front of a TV can still see 3D. In the demonstrations at trade shows, you literally have to stand over footprints that are placed on the ground to see the 3D effect, and there are only three or four places you can stand to make it work.

Glasses-free 3D TV will become a reality, but it's probably three to five years out. Like all consumer technologies, it's bound to be much more expensive than the standard 3D systems initially, so the companies can make back their development money.

4K TVs: Higher than High Definition

For years, high-definition TVs (or HDTVs) were presented as the pinnacle of sharpness. But in fact, there is no limit as to how sharp a picture can be.

3D TV seems to have attracted little interest (in Britain, the BBC ended its 3D broadcasts due to lack of viewers), so the consumer electronics industry is looking for something else to boost their sales.

They hope it will be 4K, or Ultra High Definition TV.

Perhaps you've occasionally arrived late for a movie and been forced to sit in the front rows. The picture usually looks much less sharp than if you were far away. But if the image had a higher resolution it could look good at a close viewing distance, even though you'd still get a stiff neck from the experience.

That's what Ultra HD does. With four times the resolution of plain old HDTV, you can stand even closer to the screen and still see a sharp image. Alternatively, you can buy a super-sized screen and watch it up close.

Should I Buy a 4K Ultra HD TV?

4K TVs are considerably more expensive than regular HDTVs, although their price has already started to decline.

But even if you can afford one, the question is whether you'll be able to see the difference in picture sharpness.

Based on the typical screen size, most of us won't.

According to Raymond Soneira, a theoretical physicist and owner of a company that evaluates video displays, if you sit 10 feet from the screen—typical for consumers—you'd need an 80-inch TV or larger before you could perceive that the image was sharper than that coming from a regular HDTV. How many people have room for an 80-inch TV?

Are There 4K Movies or TV Shows Available?

The other problem with 4K technology is that there is currently no way to get programs that have been shot in 4K. Blu-ray discs cannot store 4K, although there is talk about a new Blu-ray disc standard that will (in which case you'll need a new Blu-ray player). And cable and satellite companies do not broadcast 4K resolution programs.

If There's No 4K Programming, What Good Is a 4K TV?

All 4K displays are able to upscale regular high definition programming. What that means is that the TV tries to increase the resolution by increasing the number of pixels in the image. But it's not the same as watching material that was shot in 4K and then displayed on a 4K TV.

Besides bragging rights, at this time there may be little point for most of us in buying a 4K TV.

4

DVRs and TiVos: Watch What You Want, When You Want

"We don't need a TiVo," my wife said. "We hardly ever watch TV."

Our TV-watching frequency hasn't changed; it's still about three hours per week. But my wife's attitude toward having a digital video recorder certainly has. Now she says, "I couldn't imagine watching TV without one."

The difference is that my wife has discovered what millions of others have: that using a TiVo, also known as a digital video recorder, fundamentally changes the nature of the TV viewing experience for the better in ways that a VCR never did.

What's the Difference Between a DVR and a TiVo?

Nothing—at least nothing substantial. TiVo popularized the concept of a digital video recorder, essentially a hard disk that records TV. But for various reasons, TiVo has lost the lion's share of the market. DVR is simply the generic term for what TiVo popularized.

What's the Difference Between a DVR and a VHS Machine?

Plenty. VHS videotape machines are *linear* devices. That is, they record programs and play them back from start to finish. If you want to skip a section, you have to fast-forward or fast-reverse the tape, running past the parts you're not interested in until you get to where you want to be.

DVRs are *digital*. Just like all hard disks, they record material randomly. That means you can skip through the boring bits without actually having to see the material you're not interested in. DVRs are constantly running. In fact, they don't even have on–off switches. When you use a DVR, the program you're viewing, even when watching live TV, is actually the program as it was just recorded on the DVR a split second earlier.

Because a DVR is always on and recording, you can stop live TV, even if you didn't ask the DVR to record the program. DVRs typically record the last thirty minutes of whatever channel they're tuned to. The oldest section gets automatically erased as the newest part gets recorded.

How Do I Set Up a DVR?

If you have a modern HDTV it's rather simple. You connect the HDMI cable to the "HDMI out" plug on the DVR and connect the other end to the "HDMI in" input on the TV. You also connect your cable TV or satellite cable into the back of the DVR. You'll see a port labeled something like "antenna in," "TV in," "cable in," or "satellite in"; this brings the signal into the DVR. If you don't have cable or satellite, then connect the cable from your television antenna into the DVR in the same place.[9]

As I'll explain in more detail later, by using an HDMI cable you can combine audio and video into one cable; previously, one needed multiple cables for video and audio.

5

Recreating the Sound of a Movie Theater in Your Home

It's ironic: Sound is invisible, yet good sound can create a great visual experience. Rich, high-fidelity voices and effects emanating from multiple speakers placed around a room bring life to the picture, making the visuals themselves appear fuller.

Just imagine an action sequence with helicopters flying overhead, bullets whizzing by, and characters shouting back and forth across a field. Would the experience be the same if all the sounds came from a tiny, tinny speaker in the front of the TV, compared to five or six speakers in front of, to the sides, and behind you?

Creating a "theater in your home" has historically been a major undertaking; for many people, it was more of a hassle than it was worth. You had to buy the speakers and the audio receiver; they were expensive and a pain to set up. Things got even worse with theater surround sound. Now you had speakers to the left, right, and behind you, with a maze of cables to match.

Buying a receiver can be pretty dizzying, and to make you even more anxious about buying the latest and greatest, receivers come with a slew of features, most of which you'll never use. Here are the gobbledy-gook selling points of a typical sub-$300 receiver found on the Amazon Web site (no reason to mention the name because most receivers offer similar technology features):

- 80 Watts per Channel at 8 Ohms, 20 Hz-20 kHz, 0.7%, 2 Channels Driven, FTC; 110 Watts per Channel at 6 Ohms, 1 kHz, 0.9%, 2 Channels Driven, FTC

- 4 HDMI Inputs and 1 Output, HDMI Support for 3D, Audio Return Channel, Deep Color, X.V.Color

- LipSync, Dolby TrueHD, DTS-HD Master Audio, DVD-Audio

- Direct Digital Connection of iPod/iPhone via Front-Panel USB Port

- Easy to Use On screen Display (OSD) via HDMI

Recreating the Sound of a Movie Theater in Your Home

Before you run off screaming, let's simplify things.

How Many Speakers Do I Need?

To get true surround sound, giving you the sense that sounds are coming from the front, sides, and rear of you, you need speakers in the front, sides, and rear of you. But how many do you need? The typical surround sound setup consists of six speakers:

- One in the front, usually placed on top of the TV. That's where a character's voice will come from.

- Two side speakers to the left and right of the center. Those deliver peripheral sounds; for example, the sound of a car traveling from left to right will travel across those speakers.

- Two rear speakers placed to the left and right, behind you, will perform the same tasks for sounds that are supposed to be, as you may have guessed, behind you.

- And one fat, boxy speaker called a subwoofer is used to create the low-frequency rumblings essential to action movies: jet engine roars, over-the-top thumping footsteps of romping dinosaurs, or a heavy kick to the jaw in Kung Fu movies.

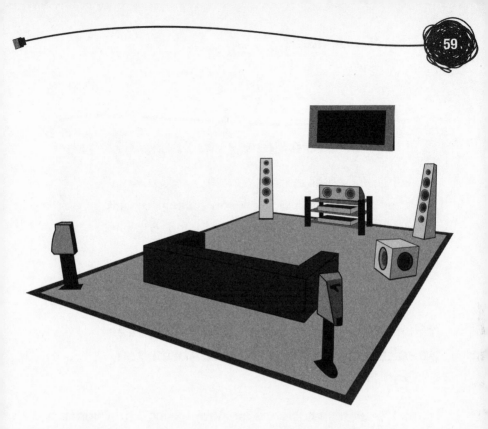

Unlike the standard speakers, a subwoofer can be placed pretty much anywhere in a room. That's because low-frequency notes are omnidirectional, so it doesn't matter where you put the speaker, as long as you can hear it.

This six-speaker combination is called, somewhat unexpectedly, 5.1. The "5" signifies five speakers, and the ".1" designates the subwoofer. The subwoofer can reproduce only a subset of the audible spectrum, just the low notes that enrich a scene. So someone decided that if a subwoofer can't play the entire range of audible sounds, it doesn't deserve a whole integer of its own.

Recreating the Sound of a Movie Theater in Your Home

Why Are Subwoofers Used?

A home theater system's speakers are so small they can't reproduce really low notes very well. Therefore, a larger subwoofer is needed to fill the gap.

If Five Speakers Are Good, Aren't Seven Speakers Even Better? How About Ten, or One Hundred?

Home theater manufacturers are now making 7.1 systems; the extra two speakers are used to fill in the gaps above your head, rather than just to one side or the other. Still, there's a law of diminishing returns. Because of the nature of sound, you don't need a speaker everywhere you want to perceive a sound. Cleverly written algorithms can make you think sounds are coming from the side even if the speaker is in the front.

6

Surround Sound Technologies

> "It has become appallingly obvious that our technology has exceeded our humanity."
>
> —*Albert Einstein*

Why Do I Want Surround Sound?

If you watch movies or TV dramas at home, they're likely to be recorded in surround sound. Using an audio system that produces sound that "surrounds" you will give you a greater sense of realism. The more authentic the sound, the more you'll be drawn into what you're watching.

I Can't Figure Out What Speakers and Audio Receiver to Buy, So What Do I Do?

The consumer electronics whizzes figured out that the home theater setup was too complex for most people, regardless of the benefits. So they came up with a new concept: home theater in a box (or HTiB, as the trade magazines call it). The idea is to sell a *solution* rather than parts: You buy the receiver and DVD or Blu-ray player already bundled with the speakers. And because everything is packaged together, connecting the speakers to the receiver is usually pretty simple; some companies even color-code the connecting cables.

To help consumers spend more of their money, the electronics manufacturers added many features to HTiB that they say enhance the surround sound experience, such as simulated listening environments that mimic different rooms. The idea is that if you close your eyes, you'll think you are in a thousand-seat concert hall, even though you are really listening in your closet-sized Manhattan living room.

I'm Not Going to Drag Cables Across the Living Room Just to Get Speakers Behind My Sofa

You're not alone, which is why you can now buy surround sound systems with wireless rear speakers. The front ones use cables to connect to the receiver, and the rear speakers connect using wireless technology.

Setting Up Wireless Speakers in the Rear Is Just Too Much for Me

There's a solution for you as well. Remember that speakers don't actually have to be in the exact position where the sound is meant to be perceived. Advanced technologies make it possible for sounds to be heard as if they are coming from the rear, even though the speaker is in front of you.

It seems illogical, but sounds appear to be coming from a particular direction not because they come from there but because various split-second delays in the time it takes sounds to reach our ears give us the distance cues we need to sense where the sounds are emanating from. In theory, you could make a noise sound as if it is coming from the rear using a speaker in the front.

In fact, you can. *Speaker bars*—a row of speakers in a box placed under the TV—can simulate sounds as if they're coming from all parts of the room. If you're short on money, space, and hassle tolerance, speaker bars are the way to go.

Why Can't I Just Use the TV's Built-In Speakers?

You can, but because TVs are getting thinner and thinner, the space devoted to speakers is getting smaller. That means the quality of the speakers built into a TV is getting pretty bad. You just spent a lot of money on a new HDTV. Do you really want to have tinny, weak sound?

Dolby Digital, Dolby HD, DTS, DTS HD, SRS, THX: What Are They, and Which Are Important?

The audio standard used on DVDs and Blu-ray discs is Dolby Digital. DTS is an optional standard for Blu-rays. They're from two separate companies that compete with each other, but the average Jane or Joe would have a tough time telling them apart. Those two are the only ones you really need to know about.

Buying hint: All home theater receivers come with Dolby Digital; most also include DTS. As long as you have Dolby Digital in the receiver, you're ready to experience surround sound.

7

Programming: How to Get It and How to Save Money

Thanks to cable and satellite, picture snow and ghosting have become things of the past. Rabbit ears and rooftop antennas are now mostly in the garbage. But cable and satellite TV has become prohibitively expensive. Spending more than $100 per month to receive many channels that you'll never watch does not always seem wise. So how can you cut the cost?

One way is to get an antenna. The same old rooftop or rabbit ears antenna can be used to receive free HDTV programming from your local networks, and because the signals are now digital, they'll be completely free of ghosts and picture snow.

You can also cancel your pay TV services and rent the shows you want. Save the $25 per month you were spending for HBO or Showtime and rent a DVD version of *Curb Your Enthusiasm* or *Weeds* on DVD from companies such as Netflix, Blockbuster, or the Redbox kiosks in grocery stores. Redbox rentals are only about $1 per night. Netflix subscriptions cost as little as $8 per month for unlimited DVD rentals, with one disc at your house at any one time. Hint: Get Netflix for a month, catch up on all your shows, and then cancel the service for a few months until you need to catch up again.

Or use the Internet and watch TV programming on your computer. You can even transfer it to your TV or get programming through a "connected TV" (more on this later).

8

Wires, Wires, and More Wires! Does This Plug into That?

"Make a product easy and simple and people will buy it. Make it complex and cumbersome and you will fail."

—*Eisuke Tsuyuzaki, former Chief Technology Officer, Panasonic North America*

They are *cables*, not *wires*. But whatever you call them, figuring out how to connect all your gear can be daunting. Look at the back of your new TV; what do you see? If you have a brand-new HDTV what you'll see is holes—lots of them! Inputs, outputs, digital connectors, analog connectors; the back of a TV looks like the telephone switchboard in a Lily Tomlin sketch. You'll find a similar dizzying array in the back of a Blu-ray player, a cable or satellite box, or an audio receiver as well.

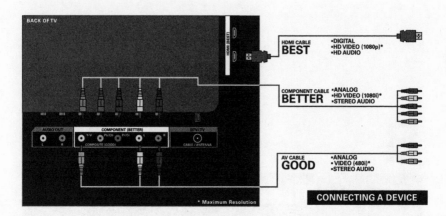

Why Are There So Many Connectors in the Back of a TV?

Connectors are constantly changing. When home computers first became popular in the 1980s, the plugs you needed to connect the PC to the printer were enormous because initially, computers were being used only by businesses. Eventually the plugs shrank, making them much more friendly to consumers. But even more than the esthetics, the new plugs allowed the devices they were connecting to do more things or do the same things faster, with better quality.

A Plug Cheat Sheet

~ *Video* ~

- **RCA (composite) cables:** two for audio (one red, one white) and one for video (yellow). Provides the worst possible picture quality between the TV and the video source (the DVD or cable/satellite box). RCA first popularized them, hence the name. Can be used only for standard definition video (not for HDTV).

- **Component cables:** better picture quality. They use three plugs for the picture, but they still need the red and white RCA cables to transmit the sound. Can be used with HDTVs.

- **HDMI:** best picture quality, plus the added benefit of being able to combine the audio and video cables into one single cable with a small plug. Only an HDMI cable can carry the highest-quality Blu-ray resolution. The Blu-ray industry made the use of HDMI a requirement for best quality; for various reasons it helps prevent unscrupulous people from making illegal copies of the disc.

~ *Audio* ~

- **RCA cables:** use the red and white cables for analog sound.

- **Digital cable:** one plug that lets you access a digital audio feed that may be available from DVD and Blu-ray players, and cable and satellite boxes.

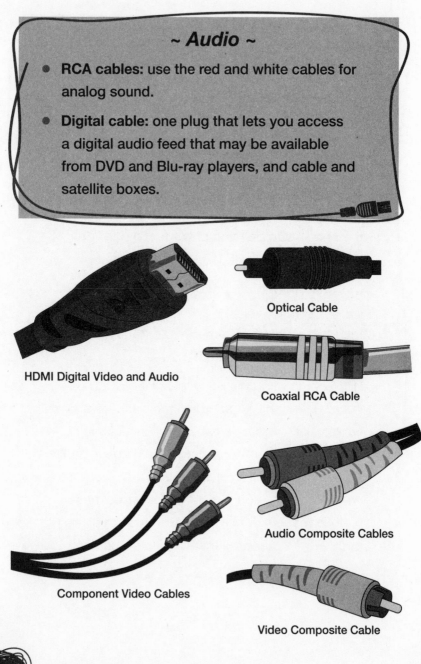

HDMI Digital Video and Audio

Optical Cable

Coaxial RCA Cable

Component Video Cables

Audio Composite Cables

Video Composite Cable

A Simple Guide to Connecting Everything Audiovisual

Except for Apple (throughout this book you can safely insert "except for Apple" for lots of things), most consumer electronics manufacturers seem incapable of writing coherent instruction manuals. They show page after page of features with little or no explanation about what those features do. The badly drawn diagrams are too small to see or too complex to understand. And the English text was written by someone for whom English seems to be a third language. For example, try to figure out what this paragraph means (taken from an actual Sony HDTV manual):

Video Color Space (x.v. color)—Displays moving pictures that is (sic) more faithful to the original source by matching the color space of source.

Setting is fixed to **Normal** when the input signal is HDMI (RGB), even if **x.v. COLOR** is selected.

In the simplest possible terms, here's all you really need to know to make the connections you need:

- **"Input" is the place on the TV to plug in the cable carrying the signal *from* the Blu-ray player, the DVD, or other device.**

- "Output" is the place on the Blu-ray player or the DVD to plug in the cable that's carrying the signal *to* your TV.

- HDMI: Use this cable for best results. You don't have to look for the words "input" and "output." Just connect either end of an HDMI cable from one device to another.

- All the other cables: Don't use them.

9

Remote Controls: Why Can't We All Just Get Along?

How many remotes do you have on your coffee table? If you've got a TV, home theater receiver, Blu-ray or DVD player, and a cable or satellite box, that's at least four. You need one to turn on the TV, another to fire up the cable, and a third to listen to the sound. If you've read the instructions, you've probably figured out how to turn on the TV with the cable or satellite box remote. But anything beyond that, such as using the same remote to run the DVD player, is either impossible or an enormous hassle.

Everyone knows this. And various companies keep on trying to come up with "universal" remotes, one remote that will work everything.

Why Can't One Remote Be Designed to Control Everything?

It's constantly tried, but the problem is that none of them work very well because there's too little space on a remote to fit all the commands that operate the various boxes. So some developers have tried to use touch screens to add lots of commands, but a command you need often may be buried three screens deep on a universal remote. I've tried at least ten universal remotes; some of them even just turn an iPad or iPhone into a remote, and none of them work satisfactorily.

Will we ever have a useful universal remote? Yes, but it will probably use a combination of voice and gesture commands to obviate the need for screens and multiple buttons altogether.

Tip: Stick to your bowlful of remotes until Apple or some other clever company comes out with a real solution.

I Got Everything Set Up, But Now My Remote Doesn't Work!

If you have cable or satellite, you probably have a remote from your programming provider that's supposed to turn on your TV, your cable or satellite box, and perhaps your audio receiver. But it doesn't. Here are a few tips to help troubleshoot the problem:

- Did you check the batteries? It sounds stupid, but dead batteries are often the problem.

- Is the slider in the right position? The carriers' remotes often have a slider that, if it's in the correct position, is designed to turn multiple devices on and off. If it's in the wrong position, such as the "TV" position, then it'll only turn on the TV.

- Did you program the remote to work with your TV as well as your cable or satellite box? You can't expect the remote to turn on other boxes if you haven't taught it what type of TV or audio receiver or DVD player you have. Programming the remote is almost always simple; you scroll through an on-screen list of manufacturers. When you find yours, you'll see a list of codes. You enter the first one into your remote (as per the cable or satellite company's directions) and try the remote. If it turns off your TV (or whatever device you're trying to get to work) you're done. If it doesn't work, you go on to the next code. Eventually, one of them will work. It always seems like it's the last one in the list.

- Are you using the wrong remote? For example, if you have a TV, a Blu-ray player, and an audio receiver all from the same company, each will come with its own remote. Often, each remote has two similar-looking power buttons: one for the device that it was shipped with and another to turn on all devices from the same manufacturer. If you push the device-specific power button, nothing will happen with the other things you're trying to turn on.

Save a Lot of Money on Cables

You've spent several thousand dollars on a new flat-panel TV and a surround sound system; shouldn't you spend just a few hundred dollars to buy those really good, professional cables from companies such as Monster Cable? After all, an HDTV is only as good as its weakest link, right?

Absolutely not.

On Amazon, you can buy an 8-foot HDMI cable from Monster Cable for $45, or you can get a 10-foot no-name HDMI cable for $2.67. Think of the profit margins!

Remember: Digital is either on or off; you either get a great picture and sound, or you get nothing. When it comes to cables, save your money and send it to Uncle Soupy.

10

Getting More from Your TV

Watching TV on Your Computer, and Vice Versa

Look at your computer screen. If it's a flat-panel model, it probably looks similar to your flat-panel TV, except that it has a keyboard. Not only does a TV look like a PC, but what you can do with each, once very different, is beginning to meld.

You can watch TV on your PC.

You can access the Internet on your TV.

Many cable channels offer various programs via the Internet. For example, you can watch BBC news clips on the BBCnews.com Web site. If you're a *Daily Show* fan, go to thedailyshow.com to see clips of previous shows.

Watch uncut movies, TV shows, and original shows on Sony's Crackle.com. It features some programming you may have forgotten (such as 1994's *The Professional*) and others you haven't (including *Seinfeld*).

Hulu (hulu.com) features TV shows from ABC, Comedy Central, Fox, NBC, and PBS. Hulu Plus, a pay version, allows subscribers to watch programming not just on a computer but also on portable devices such as smartphones, tablets, and Internet-connected TVs. Hulu Plus videos are offered in high definition; regular Hulu shows programs in standard definition. Hulu Plus subscribers can see many more episodes of a series; standard no-charge Hulu typically offers only the five most recent episodes.

Missed *Good Morning America, CBS This Morning,* or the *Today* show? Each makes clips available on their respective Web sites.

If you subscribe to HBO through most cable or satellite services, you can watch previously aired HBO programming using their HBO Go service on your PC, tablet, and some Internet-connected TVs.

YouTube became the first popular video-sharing Web site. Anyone can download anything to appear on the site within YouTube's content rules—that means no nudity—and apparently most of the world has. But YouTube is much more than cats playing pianos. It has become the world's repository for essential, historic, and ground-breaking moving images.

Hint: Think of YouTube as Your College's Library Stacks

As a student, I loved to wander down my university library's stacks, picking up a volume here or there, allowing my interests to wander among a wide variety of subjects I never before considered. YouTube gives you the ability to do the same in the visual realm.

Interested in 1980s TV commercials for the Renault Le Car, or the elusive Ipana toothpaste commercial starring Bucky Beaver? Looking for the last public presentation given by Steve Jobs, or a video showing how Margaret Thatcher learned to change her manner of speaking? What once would have taken weeks of research is now available instantly on YouTube.

Is There Any Way to Watch Programming That Doesn't Originate from Where I Live?

Television programs from around the world are now available to be streamed on the Internet. However, that doesn't mean you'll be able to watch them.

The transmission rights to certain programs are often restricted based on the country in which you live. So you may be able to watch the local BBC TV news show on the Internet if you live in the United Kingdom, but you can't do so in the United States. That's because BBC news is distributed in the States by BBC America and PBS, and neither channel wants to see its viewers siphoned off from their own networks.

Every computer connected to the Internet has its unique IP, or Internet Protocol address. The Web servers that store programming check your IP address, and if it's emanating from a country where the Internet broadcaster doesn't have permission to stream the show, you won't be able to see it.

There are workarounds, ways to spoof the location of your IP address so a computer server thinks it's coming from an authorized country. But given that it skirts legality, I'll leave it up to you (or your child) to figure out how to do it.

So What's the Downside?

Do you want to watch TV on your PC? It depends on the size of your screen. If you have a computer monitor the size of your living room TV, there's no problem. But if your computer is much smaller and the movie you're watching occupies only half the screen, then you may not like it much. (Unless you're fifteen years old and size matters little as long as you can watch anything at any time.)

The solution is to transfer the Internet image that's on your PC to your TV screen. If you have a newer computer you can probably use a DVI-HDMI cable. Older computers use another variation, such as an S-video cable.

If that sounds like too much of a hassle, and you're happy watching video downloaded from the Internet on your PC, the next section is not for you. But if you want to watch TV programming on a TV even if it comes from the Internet, there are new, simpler ways to do it.

11

Internet-Connected TVs

"We lack common areas, a time when everyone watched the same three TV channels. Today, things are fragmented, and that gives the individual too much control. You no longer have the serendipity of finding things [the way you can] when reading articles in a print newspaper."

—*Steve Mayer*

Today, most TVs can connect to the Internet. They do this for multiple reasons:

- To access the Internet (obviously).

- To download the latest software that keeps your digital TV up to date.

- To turn your TV into a giant Skype viewer (Skype lets you make free voice and video calls to other Skype users anywhere in the world).

- To use Internet apps, software that lets you watch Internet-based programming.

Television manufacturers are offering their own assortment of connected TV apps. Some, such as Amazon Video-on-Demand, Netflix, and Major League Baseball, are found on most TVs, but others may be exclusive to one brand (e.g., Panasonic's app that lets you transfer your weight readings from a special bathroom scale to the TV screen). Some Internet-connected TVs allow users to access the Web only through their apps; others allow viewers to actually go to any Web site using a browser, just like on a computer. There are Internet-connected TVs that use a standard TV remote, and others that use a remote that has a miniature computer keyboard.

Buy an Internet-connected TV, but if you're not ready for that, buy an Internet-connected Blu-ray player for $100 or less. Many of the same apps—such as Netflix—can be accessed through the player. There are also other alternatives:

- **Apple TV: This $99 box that's not really a TV lets viewers watch TV shows and movies purchased from iTunes on their TVs. You can also use it to stream movies from your computer (or movies you access from Web sites on your computer) onto your TV screen. You can even watch Britain's Sky News all-news channel. It also plays radio stations accessed through the Internet.**

Internet-Connected TVs

- **Boxee:** For about $170, the Boxee box (boxee.tv) lets you access any movies or TV shows you can find on the Internet, plus movies and TV shows from Netflix (subscription required), Vudu (see below), Major League Baseball, stuff stored on your own computer, music from the Pandora service, and photos stored on Flickr and Facebook. No monthly fee.

- **Vudu:** The pay-per-view service provides access to movies on different devices such as HDTVs, Blu-ray players, or video game consoles such as the Xbox or PlayStation, or on your computer. The price for each movie depends on the resolution you want for viewing the film. For example, as of this writing, a standard-definition film typically costs $4 to rent for two days; watching it in the highest resolution of high definition (only on a TV) costs $6. You can also buy movies; for example, the 3D version of *Toy Story 3* costs $22. Whether you buy or rent from Vudu, the movies are downloaded or streamed onto your TV, computer, or Blu-ray player. No physical discs are sent to you.

Are We Drowning in Distractions?

Are Web sites such as Facebook, Pinterest, and Twitter bad for us? Are we spending too much time in virtual worlds and forgetting how to communicate? We might like our children to communicate face to face, but as Rabbi Mordecai Finley, head of a prominent Los Angeles congregation, says, "Six kids in a room won't happen. They're so overscheduled. I don't notice any impediment in sociability. Are you a worse person if you text? I don't see it. All I ask is that [they] be present. I tell my kids, 'If you want to text, tell us, and then don't be present.'"

Internet-Connected TVs

12

Creating a Home Network

> "Today, everything wants to be connected.
> What's truly amazing is that it works at all."
>
> —*Steve Mayer*

The computer industry wants us to think that most Americans are streaming Netflix movies, Twitter feeds, photos, and music to every TV in the house while simultaneously surfing the Web.

That's nonsense.

When the instructions for one popular router—the product that's needed to distribute an Internet connection throughout the house—asks the average consumer to choose between the WPA-PSK (TKIP) and the WPA-PSK (TKIP) + WPA2-PSK (AES) methods of security, you know something's wrong.

Do you even need a home network? You do if you want to access the Internet wirelessly throughout your house using your PC, smartphone, iPad, Blu-ray player, or even—now get this—your refrigerator or oven (it's happening today). Why would you want to connect your refrigerator or oven to the Internet? Because soon a repair person will be able to figure out what's wrong with your appliances by going to a Web site and diagnosing the problem. Imagine the day when you can turn on your oven while you're still on your way home by going to a Web site on your smartphone—that day is right around the corner.

Connecting your TV and your Blu-ray player to the Internet makes sense because a lot of new programming such as Netflix movies can be streamed to your TV. And Blu-ray discs have additional, hidden information that can be accessed only if the player is connected to the Web.

The simplest way to get it all connected is to create a wireless network in your home. Unfortunately, figuring out which home networking gear you need to do all these fun and nefarious things is like buying toothpaste: There are too many choices, wild performance claims, and no idea what you need for best results.

You'll need a wireless router to get an Internet signal from your modem to a PC at the other end of the house. Routers also act as firewalls, helping to keep out malicious hackers, so you should have one, even if you have only one computer online.

Buy a router.

What Kind of Router Should I Buy?

Routers typically use either 802.11g, 802.11n, or 802.11ac technology. The AC technology signals travel further and faster.

Get an 802.11ac router.

How Much Should I Spend?

As little as possible.

With wireless N routers available for as little as $20 and 802.11ac routers at around $100, you can always buy another if it breaks. All routers use the same hardware; where they differ is in their software. So spend as little as possible once you know the features you want.

What Router Speed Do I Need?

N wireless routers transmit signals up to 300 megabits per second, way faster than a typical Internet signal, which typically is 5–7 megabits per second. AC routers can theoretically transmit at twice the speed of N routers.

Sounds like there's a disconnect here. The high router transmission speed is only the speed at which the transmission starts. Once that signal has traveled through your house, it may have lost as much as 70 percent of its speed. So the higher the starting speed of a router, the more it'll have left by the time it gets to its destination.

Higher-speed routers also allow you to transfer files faster within a house, for example, if you wanted to send a movie from your PC to your TV. They also allow you to back up files faster.

So any N or AC router will probably be fast enough.

Dual vs. Single Band

N and AC routers carry the signal on either the 2.4-megahertz band or on both the 2.4-megahertz and 5-megahertz bands. Microwave ovens and cordless phones also use the 2.4-megahertz band, so a router placed too close to those devices could lose its signal; if that happens, you'll have to restart your modem and router. The 5-megahertz band is resistant to microwaves and other appliances. But the 5-megahertz signal does not travel as far.

So here's what to do: If your router is in the kitchen near vacuum cleaners or cordless phones, get a dual-band router.

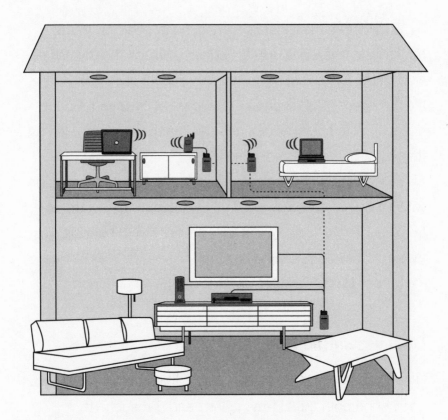

Wired vs. Wireless: Nobody's Perfect

Wireless is certainly convenient—you get an Internet signal everywhere—but there are still problems. For example, you may occasionally lose your signal, and wireless connections are typically 30 percent slower. So use a wired connection whenever possible.

If using wires is impractical (e.g., over a long distance) another solution is to use Powerline networking (also known as HomePlug). Powerline is a technology that sends Internet signals piggybacked onto your home's electrical wiring. To make Powerline work, you buy two Powerline transmitters (available from Belkin, Netgear, and other manufacturers), little boxes that simply plug into wall outlets. In each box is a plug to insert an Ethernet cable. You plug one box into an outlet near your router and connect an Ethernet cable from the router into the Powerline module. Simple. The other Powerline box is plugged into the wall where you want the Internet signal. Just like the first, you connect an Ethernet cable to this box and connect the other end of the cable to the device that needs an Internet connection.

Unfortunately, Powerline technology comes with its own set of problems. For one, be prepared to lose your signal. In Powerline tests using different products, I have lost the Internet connection every few weeks; often, restarting the PC, the router, and the Powerline modules does nothing. For some unknown reason, only an overnight hiatus brings the signal back.

When possible, create a wired, not a wireless network. Or bridge the wireless gap with Powerline technology, but still keep a wireless connection as backup for when your Powerline system goes down.

Creating a Home Network

How Do I Get the Internet into My TV, My Blu-ray Player, and My Xbox 360, All at Once?

Some TVs and Blu-ray players accept wireless signals, and Microsoft sells a wireless adapter for its Xbox 360. In those cases, it's easy to get the Internet into several places because if you're receiving a signal wirelessly, it can go to multiple devices at the same time.

If your products can't receive a wireless signal, you'll need to connect each using an Ethernet cable. To do that, bring the signal close to your entertainment products using a Powerline module, and then connect a small box called an Ethernet switch to the Powerline module. This little box is like an electrical extension cord plug expander. You plug one Ethernet cable from the Powerline module into it, and the other side has multiple holes where you can plug in up to about five Ethernet cables to connect the box to the video game machine, the Blu-ray player, and so on.

Ethernet switches are cheap. You can pick one up for $25.

If you want additional information or you still can't figure it out, there are multiple sources where you can learn more.

Amazon (amzn.to/hA7HPf), Best Buy (bit.ly/eOEa6G), and CNET (cnet.co/g4E4r0) offer comprehensive home networking primers. If you prefer to talk to a live person, visit a big-box store such as Costco or Best Buy. These days, they actually know what they're talking about.

Wireless Network Cheat Sheet

Buy an N or AC router.

Buy the cheapest N or AC router that has the features you want and that you can afford.

Get a single-band model. Exception: If you're putting your router near a microwave oven or cordless phone, buy a dual-band router.

A wired Powerline AV connection is faster and more reliable than a wireless one. Use it, but keep wireless as a backup.

13

Should I Get Rid of My Landline Phone?

> "People have set points of sociability and interaction that they need. Change the field, and they'll find other ways of interacting."
>
> —*Steve Mayer*

With cell phones ubiquitous and smartphones nearly so, is there any reason to have a landline phone any more? Young people are increasingly relying on their mobile phones as their sole source of contact. After all, why bother paying for a landline phone, tied as it is to a physical place, when you've got a cell phone that can find you just about anywhere in the world? I can think of a few reasons to keep a landline phone:

- Cell phone voice quality—especially in the United States—stinks.

- Cell phone reception is spotty. You may get decent coverage in front of your home or down the street, but inside your home can be a completely different story.

- A landline phone's voice quality is far superior to a cell phone's. With a landline you can hear the people you're speaking to very well. In fact, AT&T used to advertise the fact that it had improved the fidelity of its landline system, offering customers an 800 number that would demonstrate the differences between its old and new technologies.

- A landline phone is tied to a physical locale. If you dial your kid, you know where she or he is. Speak to someone on a cell phone, and they might be in Timbuktu (literally).

- A landline phone acts as a social adhesive. A call can be for anyone in the family, and everyone can jump on the phone to chat with friends or relatives. A cell phone is solitary; the call that comes in is meant for that person alone, and there's no easy way to share the conversation.

Should I Get Rid of My Landline Phone?

Money-Saving Tip

Phone, satellite TV, and cable companies operate under the philosophy of "if they don't ask, we won't give." When your introductory offer with one of them is about to expire, ask the company for a new deal. Tell the operator you want to speak to their "retention specialist"; threaten to quit unless they extend or lower your current price.

For example, AT&T's Internet access price is advertised as the rate for the first year only. When it was about to expire, I called and received an even better rate than I started with. If I hadn't asked, my rate would have doubled.

The satellite company DirecTV will typically give you many free months of HBO or Showtime and knock about $100 off your annual rate just to get you to commit to two more years. And why not do it? Assuming you're still alive, you'll want TV in two years. Why does this work? Because the phone and entertainment companies are in a constant battle to keep every subscriber they can. If they refuse, you've lost nothing, and you can switch to another provider anyway.

The chances of that happening are slim.

The good news is that landlines are cheap. As fewer people keep them, the phone companies offer great deals to retain subscribers. Twenty years ago, a coast-to-coast call cost 25¢ per minute during the day. Today you can pay as little as $30 per month for unlimited calling anywhere in the United States, plus caller ID, three-way calling, and many other features that you'll never use or even remember.

What Alternatives Are There to Landline Phones?

If you don't like AT&T, Verizon, or whoever serves your area, you can also subscribe to voice over IP (VoIP) service. VoIP uses the Internet, rather than the standard phone lines, to connect your calls; two of the many companies that offer VoIP service are Ooma and Vonage.

VoIP service can be much cheaper than standard phone service. For example, Vonage currently charges $26 per month for unlimited calling in the United States and Canada and to the landlines in fifty-eight other countries.

Ooma charges no monthly fee when you buy their gateway box, a device that goes between your Internet connection and your phone. Once you have it, you can make unlimited calls in the United States and Canada. Calls to other countries can be as little as 1¢ per minute if you buy a monthly bundle of minutes, or more if you just pay on a per-call basis.

If you want a lot of other features, such as an automatic second line and the ability to block telemarketing calls from certain numbers, then you have to pay a monthly fee.

Still, VoIP may not be for you.

VoIP's sound quality is often inferior to standard landline service, and if you have a home security system, it may not work with a VoIP service. In fact, Vonage recommends that subscribers with alarm systems keep basic standard landline phone service. If you have to do that, the savings may be gone.

Think about where you call. Although the idea of free calling to many countries sounds cool, if you rarely call them, what's the big deal? And with any VoIP service, you'll still need to pay for an Internet connection. Often, the big companies such as AT&T that provide Internet service charge more if you don't also get phone service from them. So add up all the costs, think about your needs, and see whether VoIP phone service still makes sense.

How Can I Call Overseas Without Spending a Fortune?

AT&T may offer unlimited rates when you call within the United States, but if you call another country without having signed up for an overseas calling plan, be prepared to faint when you get your bill. Even a call to Canada—if you haven't signed up for Canada calling—can easily run you $10 or $20 for a few minutes. Crazy. AT&T actually charges you a monthly fee for the privilege of getting an international calling plan, even if you never call a single country outside the United States.

Don't waste your money. Instead, use one of the simple dial-around plans that really do cost a lot less. Years ago, these dial-around plans were unwieldy. You had to first dial an 800 number or a five-digit code to get to the other company's service. Today, you can set up your home phone to dial the dial-around service with just two buttons; then the service can recognize your number. If you have several overseas numbers you regularly call, you can set up speed dial numbers to call them quickly.

One of the easiest and cheapest overseas call services is Pingo (pingo.com). You make the Pingo access number one of your speed dial numbers. As soon as you hear the Pingo prompt, you enter the overseas number and you're done. Rates are excellent: 1.5¢ per minute to call France and the United Kingdom, 1.75¢ to Canada, 2.25¢ per minute to call Australia and Italy, and 3.3¢ to call New Zealand. Mobile rates are higher, except to Canada.

Why Does It Cost So Much to Call Some Other Countries?

Phone rates are set by the countries that receive the calls; with communications so cheap, rates have nothing to do with distance. In most countries outside North America, there's no charge for incoming cell phone calls, so someone has to pick up the tab, and that tab can be costly. Calling a cell phone in the United Kingdom using Pingo costs 15¢ per minute—*ten times* its landline calling rate!

Countries that are short of hard currency also charge more to call from overseas. Even though it's only 90 miles away from the United States, a call to a landline in Cuba using Pingo and other similar services costs about 95¢ per minute.

Have any pals in North Korea? You'll be charged 55¢ per minute to speak to them. Once they defect to South Korea it'll cost you less than one twentieth the rate—2.5¢ per minute—to have a chat.

One Phone Company to Forget

You may have heard many commercials for PennyTalk, another dial-around service that advertises its low, low, low rate of just 1¢ per minute to call Canada and the United Kingdom. What they don't say so loudly is that every call—no matter how long—incurs a 49¢ "connection fee." So that 1¢ one-minute call to England actually costs you 50¢, one of the highest rates around. If you speak for fifteen minutes to the United Kingdom, PennyTalk will charge you 64¢, compared with Pingo's 22.5¢. Even an hour's call is more expensive: $1.09 for PennyTalk vs. 90¢ on Pingo.

The best advice: Stay away from PennyTalk. A better name for it would be Lots of Pennies to Talk.

14

Everything Is Connected

Just because you can do something technologically, should you? The same question that confronted developers of the atomic and hydrogen bombs has become a consideration for today's consumer electronics pioneers, although with fewer dire implications.

You may be able to combine a food processor with a word processor, but would it make any sense? Simply because you can synchronize a Web site with a device that ejects appropriate smells when you visit that site, should you? In the late 1990s, the people behind DigiScents, a startup with $20 million in funding and seventy employees, apparently thought it was a no-brainer. Some of us thought the people who came up with this idea had no brains, and in *The New York Times*, I referred to it as one of the dopiest tech inventions ever. In 2001, it shut down.

Just because you can connect the Internet to everything, should you? Why not connect the fridge and the oven? While we're at it, how about the scale? A clock radio? And of course, the washer–dryer.

What Can I Do with Connected Devices?

It all sounds pretty silly, but if you think about it, it makes sense. When an appliance is connected to the Internet you can

- **Find out if your parent with heart failure just gained 3 pounds overnight or suddenly went into diabetic shock. You don't need to worry anymore that your mother isn't bothering to tell you because she doesn't want to worry you. Instead, you can arrange for medical information to be sent automatically over the Internet to a doctor or caretaker.**

- **Remotely diagnose what's wrong with your fridge or washer–dryer instead of asking for a repair tech to be sent out.**

- **Automatically order parts.**

- Monitor food consumption and automatically compile a shopping list.

- Remotely control the oven, turning it on and off before you get home.

- Post your weight on a Web site for remote access anywhere in the world.

- Suggest recipes based on what's in your fridge.

- Turn on the lights or change the lights' colors while away from home, by using a smartphone app.

● ●

All these features are available today. Companies such as Bosch with its Health Buddy System, Honeywell's HomMed products, and Ideal Life are equipping thousands of patients with automatic health monitoring devices. Because doctors, nurses, and adult caregivers don't need to constantly check up on them, older adults can remain in their homes longer. Face it; given the option, that's where we would all like to spend our last days.

LG and Samsung make Internet-connected fridges, and Samsung has one with a built-in tablet and apps. You can listen to music from Pandora while you're putting away the eggs. Or if you don't have eggs, search for a quick egg-free recipe using the Epicurious app. If you can't find anything, and you simply must buy eggs, check the weather app before you run out to the store.

In the near future you'll be able to scan grocery receipts so the fridge automatically knows when products will expire, although that may seem like more effort than it's worth. And some appliances will talk to each other, so you'll be able to turn on your connected washing machine from your connected fridge.

Is There Any Reason I Wouldn't Want Everything Connected?

There is a downside. *You'll* be able to do all those things, but so will your mischievous children, a jilted lover, or a disgruntled ex–business partner. Imagine turning on your oven when you're on the 5:18 to Stamford. Then imagine someone who really dislikes you doing the same thing at noon, and turning the temperature up to 500 degrees. Of course, there'll be privacy controls to prevent malicious behavior, but one burned-down house could screw up the whole enterprise.

15

You (Used to) Light Up My Life: The New Light Bulbs That Will Change the World— and Yours

With all the rapid technological changes that have become a part of our lives, there's one technology that has been glacially slow to change: the light bulb. The standard incandescent light bulb is essentially the same as the one Thomas Edison invented more than a hundred years ago. The concept is identical and the execution not much different: Electricity is fed to a wire (filament) that glows. The interior of the lamp in which the filament is placed is a vacuum, so the filament lasts longer. The lamp screws into a socket called, appropriately enough, an Edison mount. If you live in Europe, the bulb is pushed in and given a tiny twist, just like putting a bayonet on a rifle (and guess what that mount is called?).

The great thing about this technology is that it's incredibly cheap: Bulbs cost about a buck. The bad thing is that despite the invention of the transistor and computer chips, they haven't changed much since 1910. Traditional light bulbs are also unbelievably inefficient. More than 90 percent of the energy used is not turned into light; it's turned into heat (that's why when you touch a bulb that's on, you get burned).

Are Standard Light Bulbs Being Banned?

"And the government would have banned Thomas Edison's light bulb. Oh yeah, Obama's regulators actually did just that."

—*Mitt Romney, March 19, 2012*[10]

"That's why I introduced the Light Bulb Freedom of Choice Act. I am so proud of that bill. . . . President Bachmann will allow you to buy any light bulb you want, in the United States of America."

—*Michele Bachmann, as seen on YouTube*[11]

Various people with little knowledge have made false statements about the future of light bulbs and, as Michele Bachmann said, promised to "allow you to buy any light bulb you want." But the purchase of regular light bulbs was never going to become illegal. What will eventually be prohibited is the *manufacture* of regular light bulbs that are highly inefficient. You can always buy whatever's on a store's shelf.

According to government guidelines, each year most incandescent light bulbs that are sold in the United States will have to use only a certain amount of power to create the same amount of light that formerly would have required more electricity. First, it's the 100-watt bulbs that are affected. Eventually, those standards will reach down to 40-watt bulbs. This is no different from setting gas mileage efficiency standards for vehicles or creating Energy Star standards for televisions, water heaters, and refrigerators.

It's not a "government takeover" of light bulbs. It's a policy designed to reduce energy use.

Bulbs Exempt from the Efficiency Standards

Here's a list of the regular light bulbs that the government has exempted from its efficiency standards. You'll probably be able to buy these forever, no matter how much they hog energy. Fear not—you can always buy plain old fridge lamps!

- Appliance lamp
- Black light lamp
- Bug lamp
- Colored lamp
- Infrared lamp
- Left-hand thread lamp
- Marine lamp
- Marine's signal service lamp
- Mine service lamp
- Plant light lamp
- Reflector lamp
- Rough service lamp
- Shatter-resistant lamp (shatterproof and shatter-protected)
- Sign service lamp
- Silver bowl lamp
- Showcase lamp
- three-way incandescent lamp
- Traffic signal lamp
- Vibration service lamp

So I'll Have to Use Those Ugly Compact Fluorescent Bulbs?

Compact fluorescent light bulbs have improved significantly. They use much less electricity, and they last quite a bit longer than regular bulbs. But when it comes to the quality of the light they create, they're still pretty dreadful. Stand under a fluorescent light, and your face takes on an ugly greenish tint. That's because fluorescent light doesn't have all the colors of the spectrum, so the light it creates doesn't look like the light we're used to seeing from regular light bulbs. And the light isn't directed; it shines in all directions, so there are no interesting shadows such as those you get from a regular light bulb. Can you imagine putting a fluorescent light in your living room or bedroom? That's essentially what you're doing when you use a compact fluorescent bulb, the kind that screws into a table lamp.

Even when a compact fluorescent lamp claims to emit warm light, it's not the same; it still looks somewhat off. There are environmental problems too; compact fluorescents contain mercury, so when you throw them away you're polluting the earth. If you live in a cold climate, don't try to use them outdoors—they often won't light. And many compact fluorescents can't be dimmed like regular bulbs.

What a dilemma! Use a regular bulb and waste a lot of energy, or use a compact fluorescent and get bad-looking light, polluting mercury, and lamps that won't dim. Fortunately there's a third alternative (you probably knew that was coming).

Are There Any New Lighting Technologies on the Horizon?

The newest technology creates light from a light-emitting diode chip: Essentially, a tiny little dot the size of a pencil point can be made to emit a ton of light. These new LED lamps use about one tenth the number of watts that a standard bulb uses to create the same amount of light. Unlike compact fluorescents, the light actually looks just about the same as that coming from a regular bulb. They can be dimmed, and they don't contain mercury.

Because the light comes from a silicon chip like those found in computers, these bulbs can do parlor tricks. You can turn them on and off from a cell phone. You can get an automatic text message if they malfunction. Combine LED chips of different colors, and you can change the color of the light at will.

Dean Kamen, the entrepreneur who invented the Segway, has converted all the lights to LEDs in his house and its surroundings on the island where he lives. You can see pictures of his house at nyti.ms/S2jiRz. LED lamps are being used to dramatically light museums and major buildings such as the Empire State Building. You can see some examples at nyti.ms/Wwxt3d.

Cities such as San Francisco are installing LED street-lights. If one burns out they'll know immediately and auto-matically, without waiting for a call from an angry resident after the street has been dark for a month. If there's no crime, half the lamps can be turned off to save energy, and if there's a riot, the lights can be strobed to disorient and disperse the troublemakers.

And here's the cool part: All that can be accomplished from a smartphone app! Because the lights use computer chips, they can be connected to the Internet and then con-trolled from any other device that's also connected, such as a computer, iPad, or smartphone.

Amazingly, LED lamps can last as long as twenty years. That's a big deal, because if you install them in your home, you won't need to change bulbs every few months, which means fewer trips to the hardware store. If you have high ceil-ings, using LED lamps means fewer climbs up a ladder and fewer potential falls off a ladder.

Because they use a fraction of the electricity used by normal bulbs, your electricity bills will go down.

Why Wouldn't I Want to Use an LED Lamp?

Right now, LED lamps cost a lot more than regular bulbs or compact fluorescents—about twenty times more. In the end, with their super-long life and tiny energy consumption, you'll still wind up saving money. But there aren't that many people willing to spend $20–30 for a light bulb, which is why they won't stay at $20–30. The world's giant lighting companies— GE, Osram Sylvania, and Philips—are spending most of their research and development money on LED research, and they're not doing this for their health. Prices will come down dramatically over the next few years, in much the same way they did with LCD and plasma flat-panel HDTVs.

How Can You Motivate People to Spend More Upfront Just for a Better Bulb?

Remember: LED lamps use computer chips to light them up. So these digital lights can easily contain additional chips to accomplish other tasks. For example, companies are now creating LED lamps with chips that can

- **Act as a booster for your Wi-Fi connection, grabbing the wireless signal and amplifying it so it reaches throughout the house.**

- Contain a speaker so sound can come out of the same spot in your ceiling as your light.

- Learn your movements and automatically shut off or turn on your lights at times when you're usually in or out of a room. That not only saves energy but also can act as a home security device.

- Control your home's temperature by sensing the heat of a room.

● ●

Should I Buy LED Lamps Now or Wait for the Price to Come Down?

If you have a bit of disposable income, buy a few LED lamps—perhaps one in the shape of a standard bulb (called an "A" bulb) that you put in a bedside lamp. If you have recessed lighting in your kitchen, bathroom, or living room, buy a substitute for one of those sizes as well. LED lamps come in all shapes and sizes: standard bulbs, multiple sizes of recessed reflector lamps, low-voltage LED substitutes (the kind you see in track lighting), chandelier lamps, under-cabinet LED strips instead of small fluorescents, and of course, Christmas lights.

Buy one every six months, and you'll probably see a noticeable drop in price each time you do.

Where Do I Get LED Lamps?

You'll find them at home improvement stores such as Home Depot and Lowe's, and of course on Amazon.com. Even Costco carries them.

If you want more information, you can read a simple explanation of LED lamps that I wrote for the *Costco Connection* magazine at bit.ly/T9TG8d.

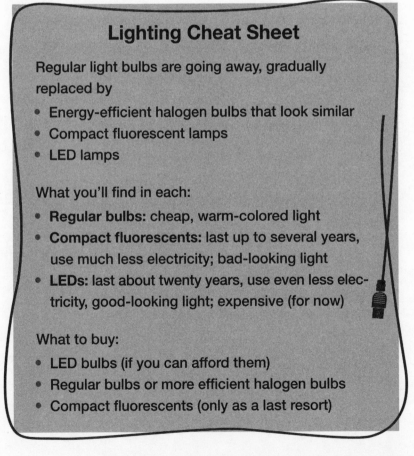

Lighting Cheat Sheet

Regular light bulbs are going away, gradually replaced by

- Energy-efficient halogen bulbs that look similar
- Compact fluorescent lamps
- LED lamps

What you'll find in each:

- **Regular bulbs:** cheap, warm-colored light
- **Compact fluorescents:** last up to several years, use much less electricity; bad-looking light
- **LEDs:** last about twenty years, use even less electricity, good-looking light; expensive (for now)

What to buy:

- LED bulbs (if you can afford them)
- Regular bulbs or more efficient halogen bulbs
- Compact fluorescents (only as a last resort)

16

Simplify Your Life Away from Home: Android, iPad, iPod, iPhone, Oy Vey!

"Every day I have a routine; I spend fifteen minutes max on the computer. I check to make sure the President's alive, and we're not at war. Then I can get to work."

—*Rabbi Mordecai Finley*

"Camera phones threaten to turn everyone into amateur paparazzi. We are witnessing our personal space shrink because of the way technology is being used."

—*Daniel Solove, Professor of Law, George Washington University*

> "The basic human needs to become connected remain the same. We can now do that far more easily, instantaneously and cheaply. The age of mass consumer electronics has evolved to the age of personal electronics where everything relevant is with you and everyone knows where you are."
>
> —*Eisuke Tsuyuzaki*

Smartphones are no longer a novelty. Until recently, they were strictly tools for businesspeople to check their e-mail on the go, but now they have become a necessity for everyone from senior citizens to seven-year-olds. Reading e-mail, finding your way with its built-in GPS, accessing the Internet—all of these things make a smartphone a must-have device. Here are some more things you can do with a smartphone:

- Talk (obviously)
- Text
- Stop your kids from texting when they're driving
- Play games
- Track your kids
- Take pictures and movies and e-mail them
- Track down your phone from a computer if it gets lost or stolen and remotely lock it so no one can use it

- Use it as a GPS to find your way, with specific directions based on whether you're driving, walking, biking, or taking the bus
- Use it as a bubble level
- Use it as a ruler
- Use it as a flashlight
- Use it as a magnifying glass
- Convert dollars to Euros and every other currency
- Convert miles per gallon to liters per 100 kilometers
- Play music, both recorded on the phone and streaming from an Internet site
- Keep track of your appointments on an electronic calendar
- Check in for a flight without printing a boarding pass
- Pay for your latte without using a credit card

What's an Android?

Android isn't a thing; it's an operating system created by Google for products such as computers, tablets, and smartphones. It's the software that allows the products to perform functions such as opening files, naming folders, sorting documents, and moving things around. Only Apple products can use Apple's operating system, but Google's Android can be used by just about any company that wants to incorporate it into their laptops, smartphones, or tablets. There aren't yet many computers that use Google, but there are plenty of smartphones and tablets that do.

Just as you can buy lots of different brands of computers that run Microsoft Windows, you can buy many different brands of smartphones (and increasingly, computers and tablets) that run Android.

How Does the Android Experience Differ from Apple's?

In many ways, they're very similar. A lot of the same gestures you use on an Apple portable product, such as pinching to make a picture bigger or smaller, are similar to what's done on an Android device. On smartphones and tablets, both Android and Apple's iOS software allow you to swipe your finger across to get from one screen to the next. Both systems use little icons representing apps, or applications. Smartphones and tablets running both systems offer thousands of apps, and most of the devices include built-in cameras, music players, and navigation functionality.

In some cases, the differences between Apple and Android functionality are subtle. For example, Apple has perfected what's called a "rubber-banding" effect. When you swipe your finger down a screen—say, down a list of e-mails—and you reach the end of the list, the last entry bounces back a bit when you remove your finger from the screen, as if a rubber band were snapping back into place. Although it might seem superfluous, the effect lets you know that you can't go any further on that screen; you don't have to wonder whether you're not swiping the screen properly. That isn't included on Android.

To get to the apps on an Android device, you have to click on the "apps" logo at the bottom of the home screen. With an Apple device, the apps screens are the first things you see.

All the preference settings for apps are in one place on Apple iOS devices: in the settings app. With an Android device, many functions are reached by touching a submenu button below the screen.

Apple extends its simplicity by using one physical button. You push it to get to your app screens, and you double-click it to activate Siri, the not-very-accurate voice navigation system. Android smartphones use three buttons, including one to pull up a submenu of options and another to take you back up one level. For example, if you've opened an e-mail in the mail app, this button will take you back up one level to your list of e-mails. And there's a center button that acts much like Apple's home button: It takes you back to the home screens, where you can see a layout of all your apps.

At times, those differences are meaningless: You may simply prefer the look of certain apps designed for the iPhone over those that work on Android. For example, both platforms include calculator apps, but for what it's worth, I think the Android version looks better. The numbers are easier to read, with a more pleasing font. However, Apple's version becomes a more sophisticated scientific calculator when the iPhone is turned horizontally; Android's doesn't.

Which Is Better? Android or Apple's iOS?

There is no "better." Apple's iOS operating system is simpler to master, but many people are very happy with their Android products. With Android, as with Microsoft's Windows system, you can choose products from many manufacturers. These may use the same operating system, but the products can differ in terms of screen size, placement of buttons, battery life, ruggedness, and so on. Once you understand the Android paradigm, it will become second nature no matter which brand you buy.

Android phones use a standard USB mini-plug to charge the phone, and Apple uses its proprietary iPhone connector. If you lose your Android cable, you can always use another that might have come with your digital camera.

If you lose your Apple cable, you'll have to buy another one.

Apple's New Proprietary Connector

Ever since the first iPhone, Apple has used the same thirty-pin connector plug, a small, wide affair. Everything that connects to an iPhone, such as car audio systems and speaker docks, has come with that same connector.

But with the new iPhone 5 and the newest generations of iPads, Apple switched to a much smaller plug the company calls Lightning. The smaller plug was one way that Apple could make the newest iPhone thinner and lighter than its predecessors. And unlike the thirty-pin plug, the Lightning plug is reversible; it connects no matter which end is up.

If you have an older speaker dock, the new iPhone 5 won't be able to plug into it, unless you buy a Lightning-to-thirty-pin adapter. Apple may eventually approve adapters from third-party companies, but as of this writing, if you want one, you have to get it from Apple.

With Android's built-in photo app, you have much more functionality than with Apple's iPhone camera app. For example, on the Android phone you can apply cartoon effects to a picture, automatically take a shot only when the subject is smiling, and crop the picture. All of this can be done on an iPhone, but it requires downloading various photo manipulation apps.

Apple's iPhone synchs seamlessly with an Apple computer. Enter an address in your iPhone's address book, and it will automatically appear on your iMac or your iPad. You can also synch your Google address book and calendar with a Mac or Windows computer, but it takes a few more steps to set up.

Whichever platform is better right now, it will soon be superseded by a better version from its competitor. As long as multiple operating systems exist, whether that's Apple's iOS, Google's Android, or Microsoft's Windows Mobile, the "better" system will constantly change hands.

What Are Apps?

Short for *applications*, apps are little programs that either provide functionality in their own right or act as convenient shortcuts to grabbing information or tools from a Web site. For example, apps with their own functionality include those that turn your smartphone into a calculator, a photo storage device, a camera, a flashlight, or a converter of weights, measures, and currency. Apps that actually access Internet information include the NPR News app, airplane reservation apps, bank account apps, newspaper apps, shopping apps such as Amazon's, the OpenTable restaurant reservation tool, and the Fandango movie timetable and ticket purchase app.

Then Why Do I Need an App If I Can Go to the Actual Web site?

You could always use a smartphone's Web browser to go to *The New York Times* Web site instead of using its iPhone app. But despite Steve Jobs's gushing excitement at the first demonstration of this feat when he introduced the original iPhone, it's virtually impossible to look at most Web sites on a smartphone unless you're a child with perfect eyesight.

Some Favorite Apps (Most Are Free)

- **Amazon PriceCheck:** Take a picture of a product's UPC barcode, and you'll quickly find out whether you can get it at Amazon for less.

- **CompareMe:** What's cheaper? Three for $1.56 or four for $2.23? This app will tell you, and by how much.

- **Flashlight:** This turns your iPhone into a flashlight and magnifying glass. Great for restaurants.

- **Google Sky Map (Android phones) and Star Walk:** Point your phone toward the sky, and these apps will overlay the constellations above you on your screen, letting you know whether that light is a star or Jupiter.

- **Multi-Convert:** This app converts anything: Fahrenheit to centigrade, U.S. dollars to Ukrainian Hryvnias, cubic decameters to U.S. dry pints, U.S. to Australian shoe sizes, miles per U.S. gallon to liters per hundred kilometers, or ergs per second to joules per minute. Crazy.

- **OpenTable:** On the road and need a reservation? Make it from your smartphone to OpenTable, and you don't have to wonder whether the bored hostess at the front desk ever bothered to write it down.

- **Square:** With the Square app and a tiny free card reader that plugs into your smartphone, you can take credit card payments.

- **Zillow:** What's your neighbor's house worth? Zillow will tell you.

QR Codes: That Little Black Box with Patterns That Appears on Many Print Ads

The box is called a QR or Quick Response code. Invented in the 1990s in Japan, it's essentially a fancy barcode. When you take a picture of the box with your smartphone (by first downloading a free QR reader app), you'll see more information directly on the screen, or you'll be directed to the advertiser's Web site to learn more. Businesses are adopting them because they think that consumers will be more likely to take a picture of a code than enter a long Web site address to get additional information about a product. Utilities are using QR codes to explain to customers why they should stop receiving

paper bills. Vehicle new car stickers have QR codes that will take you to the manufacturer's Web site and the government's fuel economy site. Many companies put codes at the bottom of magazine ads so consumers can grab their smartphones or tablets and get more information.

Both Apple's iTunes store and Android's Play Store offer many free QR readers. They all work pretty much the same and are simple to use. You open the app, touch the picture-taking button, and focus the screen on the QR code until the code locks into place; you'll know when that happens, as typically a square forms around the code. Then you're done.

Aren't Smartphones Very Expensive?

So you've decided to buy a smartphone. If you're not careful, that can cost you $100 or more per month in fees. Here's how to keep your costs down.

Think about how many minutes of calls you make per month, and get a plan that covers that amount with a bit to spare. You don't want to buy a plan with many more minutes than you need, and you also don't want to get one that comes up short. If you go over your allotted minutes, the mobile company will ding you for crazy amounts of money; it can be as much as 20¢ per minute once you've exceeded your plan.

Do you text? If you text infrequently, buying a text plan may be a waste of money. Although individual texts without a plan cost an obscene amount of money (20¢ per text, although it costs the company fractions of a penny), it may still be cheaper to pay as you go.

Block That Text Spam!

Fed up with those obnoxious telemarketing calls at all hours? Now you can get fed up with the same thing on your smartphone. Unwanted text messages—text spam—have invaded cell phones. Not only is it annoying, but unless you have unlimited texting, you're paying for every one of those incoming entreaties to get you to borrow money or sign up for a cheap (usually bogus) mortgage. They're illegal, of course, but that doesn't stop the spammers. Here's how to stop spam texts:

- **AT&T subscribers:** Pay $5 per month for Smart Limits for Wireless to selectively block texts from senders you don't want.

- **Sprint:** Selectively block texts for free by sending certain commands to "9999." Go to bit.ly/xL1USV for a list of commands.

- **T-Mobile:** Use the E-Mail and Text Tools section of the Web site to set up filters that will selectively block texts.

- **Verizon:** Go to the Spam Controls section of the company's Web site for instructions on how to selectively block texts for free: bit.ly/y0aBLV.
- **What not to do:** *Don't* text "Stop" to try and curtail text spams. That only tells the spammers your cell number is legit.

Watch your data use. If you watch lots of movies on your smartphone or tablet, that can eat up your allotted data bits and cost you higher fees once you've exceeded your amount.

How Much Data Do I Need?

If you use a smartphone essentially to send e-mail, you'll use very little data. However, if you plan to view high-definition movies on your iPhone or Android device, watch out: You'll quickly run out of data or be paying overages. A two-hour HD movie can eat up 3–5 gigabytes of data, whereas sending thousands of e-mails will use just a fraction of that amount.

To figure out what size data plan you need, use the handy AT&T calculator at att.com/att/datacalculator/. You estimate how many e-mails, movies, and songs you'll devour, and the calculator tells you what size data plan you probably need.

You can also buy a family plan; getting multiple phones on one plan will save money.

Consider mobile plans with rollover minutes. Carriers such as AT&T let you roll over unused voice minutes to the next month's allotment. Unused text messages do not roll over.

iPhone Texting Tip

If you're an iPhone user, you can text other iPhone users who are running the latest operating system, for free. With Messages, Apple's texting app, a text message is sent to fellow iPhone users as a free iMessage, using the cellular network rather than the data network. If your text message shows up in blue, it's going out over the free iMessage technology. If it's green, you're paying for it as part of your texting plan, or you'll pay for it on a per-text basis. Even texts to fellow iMessage users may occasionally show up in green. That means that the iMessage network is temporarily unavailable.

Can I Use My Mobile Phone Overseas?

When you use your mobile phone in another city in the United States, you almost always pay the same as when you're home. That's because the mobile carriers have worked out roaming deals with carriers throughout the country.

Mobile carriers have also worked out roaming deals with companies in other countries, but they're not very advantageous to the users. If you take your iPhone or Android phone to another country, you'll see another company's name where you normally see the familiar AT&T, Sprint, or Verizon name on your screen. Jump on the Orange network in France, the Virgin Mobile network in the United Kingdom, or any other foreign carrier, and you'll probably pay a fortune when you get home.

International cell phone usage is not for the faint of heart. Using my iPhone for one week in Israel to access e-mail, listen to NPR, read a few newspaper apps, and look for my hotel with Google Maps rang up a $700 bill. Fortunately, I didn't have to pony up the big bucks; Apple gave me free overseas access to check the costs for a *New York Times* article I was writing. But most users won't have that advantage.

There are alternatives, but to understand them, we need to pause for a quick lesson on cell phone technology. The United States uses two different and incompatible cell phone technologies: CDMA and GSM.[12] Europe and most of the rest of the world use GSM; GSM devices typically store some of your phone's information, including its number and your contacts, on a tiny little SIM card.[13]

China uses mostly CDMA, the system used by Verizon in the United States. These phones don't use SIM cards; all the phone's info is stored on the device. Take a CDMA phone to Europe, and it'll be useless.

Ergo, If You Take a GSM Phone from the United States and Bring It to Europe, It'll Work, Right?

Maybe. Our GSM system used by AT&T and T-Mobile operates on different frequencies from the European systems (and New Zealand's, if you're interested). So you'll need a GSM phone that also works on at least one of the frequencies used in the country you are visiting. That's a lot easier than it used to be. For instance, the iPhone is a quad-frequency phone, compatible with European standards. And many other phones today operate on the frequencies used in other countries.

So If You're an AT&T or T-Mobile Customer, the Problem Is Solved, Right?

Wrong. There's still that pesky little problem of costs. Your phone may work, but using it could cost you an arm and a leg. To keep the costs down, you can buy a local SIM card. This will give you your own number in England or Italy or wherever you are, and you'll pay much lower rates.

Hint: If you buy a local SIM card, make sure your friends back home have an international calling plan. If they don't, they'll pay up to $1 per minute to talk to you, and they may stop being your friends.

So, we've solved the international calling problem.

Not really. Even if you have a GSM phone with the right frequencies, you may not be able to use an overseas SIM card. That's because American carriers typically lock their phones electronically, preventing them from being used by other carriers. If you have a locked phone (and most smart-phones are locked) from AT&T, it won't work on T-Mobile's network, even though the SIM card from one will actually fit in the other's slot. The same thing holds true if you have a T-Mobile phone and you try to use it on Italy's TIM cellular network or Vodafone in the United Kingdom.

Why are the cellular companies such meanies? Here's what they're thinking: "Hey buddy, we gave you a great deal, charging you only $200 for a smartphone that actually cost us $600. Do you think we're crazy enough to let you use it on one of our competitors' networks and lose all that extra income we can make by charging you outrageous roaming charges?"

So what you need to do is unlock your phone. T-Mobile will unlock your phone after four months of service, and

AT&T will unlock your phone after your contract is up (but not before); to get that done, call their customer service line, but do so *before* you buy a new phone and start a new contract. Once you register a new device, the cellular provider may decide not to unlock the old one. If you have a GSM iPhone that's still under contract, you can unlock it yourself. If you know someone whose idea of a fun time when he was a kid was to take a telephone apart to see what made it work, direct him or her to www.iphonehacks.com/ultrasn0w or blog.iphone-dev.org for a free but pretty complex unlocking method. The danger with these unofficial unlocking methods is that if you don't really know what you're doing, you could damage your phone.

iPhone owners can get their devices easily unlocked at chronicunlocks.com. The fee is based on which carrier you use. Unlocking occurs automatically a few hours after you pay the fee.

Android users can try unlockandroid.net (for some Samsung models) or android-unlock.net. But if you're the kind of person who needs this book, I suspect you won't be tempted to try it.

But What If I Can't Unlock My Phone, or It Won't Work in the Country I'm in?

So let's say you're stuck with a locked phone or one that won't work in the country you're visiting. Here are a few more alternatives:

- **Purchase a reduced-price international calling plan from your carrier**

- **Rent an unlocked phone temporarily from your carrier**

- **Buy or borrow a cheap, unlocked GSM phone on eBay**

- AT&T, Sprint, and Verizon offer reduced-price overseas calling plans. If you check your e-mail mostly using a free Wifi hotspot and make just a few minutes of calls, you can still occasionally check your e-mail while out and about abroad, and pay about $100 for a week's worth of data. Alternatively, when you get to the country you're visiting and if you speak their language, buy a cheap pay-as-you-go phone from a local phone shop. Many big airports have cellular shops in the terminal. You can then pay much lower rates for local calls and top up your available balance when you're running low on minutes.

- Use Skype, either from your computer or your smartphone. Whenever you're in an Internet hotspot (they're pretty easy to find in most major hotels and at coffee houses like Starbucks), you can make free calls to people who have signed up for Skype. If you're using a laptop to do the calling and it's got a built-in Web cam, you can see them as well. Skype actually has become the best way to communicate with friends and family around the world.

● ●

17

Tablets: Android and Apple

"Another former Apple executive who was there at the time said the tablets kept getting shelved at Apple because Mr. Jobs, whose incisive critiques are often memorable, asked, in essence, what they were good for besides surfing the Web in the bathroom."

— *"Just a Touch Away, the Elusive Table PC,"*
New York Times, *October 4, 2009*

Apple's first portable computer, the Mac Portable, weighed almost sixteen pounds, had an optional 40-megabyte hard drive, and cost $6,500.

The newest iPad weighs less than one and a half pounds, comes with at least 16 gigabytes of storage (almost 320 times as much as the first portable), is at least a zillion times faster at processing data than the portable, and costs as little as $500.

It's also captivated the world's imagination. In the first three days after the 2012 iPad went on sale, Apple sold three million around the world, three times more than the first iPad during its initial weekend. When the first iPad went on sale, a popular YouTube video showed an ecstatic hundred-year-old woman using it to access the Internet for the first time.[14]

The iPad paradigm of pushing objects around on the screen, pulling two fingers apart to make them bigger, and squeezing to shrink them has become part of the popular culture. Toddlers know how to operate iPhones and assume that they can see photos on any object they pick up simply by swiping across the screen.

But iPads are not the same as computers, and Amazon's Kindle book reader (known as an e-Reader) is not the same as an iPad.

What's the Difference Between a Kindle, a Nook, and an iPad?

Amazon's Kindle models and Barnes and Noble's Nooks are designed primarily for reading. They use a technology that is easy to read in bright sunlight and uses very little battery power. You may not need to charge a Kindle more than once a month.

On the other hand, Apple's iPad is designed primarily for everything else and can also be used for reading. It creates an image using an LCD screen that's tough to read in bright light (don't even think about taking it to the beach), and it may need recharging every day or so if you use it frequently. The iPad has a bigger screen than many Kindle models; the color display makes it ideal to go on the Web, take and view photos and movies, and play games.

What Can't You Do with an iPad or Any Other Kind of Tablet?

Have you been to Disneyland or Disney World? If you have, you know that unlike other amusement parks, it's impossible to see the inner workings of the place.

You can't peek behind the stores on Main Street, or you'd see all the scaffolding that holds up the storefronts, and the

Tablets: Android and Apple

illusion of this impossibly clean and tranquil town would disappear. And it would be messy; you'd see all the pipes and electrical cables, fallen nails, and rat turds that are part of any amusement park.

The iPad—and actually all computing tablets—is designed to be like Disney's Main Street. You can't see the structure holding everything up. You can't tinker with the operating system, throw away files that may be messing with its smooth operation, or tweak the way it works. You can't peek behind the façade the tablet makers created.[15] And because tablets don't have physical keyboards, you can't type on them as easily as you can on a real computer.

Why Would I Want a Tablet, Whether It's an iPad or an Android-Based Device?

- **To carry a lightweight computing device instead of a heavier laptop**
- **To read books**
- **To have a big screen to show off your photos**
- **Because you don't like a physical keyboard**
- **To avoid dealing with computer viruses and constant system freezes**

- To easily play video games on a device that's easy to carry onto a plane

- Because you like a lot of the apps on your iPhone, and you want to use them on a larger screen

- To watch movies and TV shows while you're traveling.

- To store movies and TV shows and transfer them wirelessly to your big-screen TV

Why Wouldn't I Want a Tablet?

- You hate typing on a virtual keyboard, and you're much faster on a real one.

- You want more control over the software you use.

- You like tinkering with the innards of your computer.

- A tablet screen is too small.

- You want to watch Flash videos (many videos are only in the Flash format, and many Web sites use Flash for animations). If you have an iPad, there's no easy way to watch Flash content.

- You can't afford one (before you give up, look for a used tablet on eBay).

Tablets: Android and Apple

18

Your Digital Life, Everywhere

> "What turns me on about the digital age, what excited me personally, is that you have closed the gap between dreaming and doing. You see, it used to be that if you wanted to make a record of a song, you needed a studio and a producer. Now, you need a laptop. . . . Imagination has been decoupled from the old constraints."
>
> —*Bono,* TED Talks, *February 2005*

Digital photos and music were once confined to a computer. If you wanted to see or listen to them on your TV or smartphone, you had to connect them with a cable to transfer the files from one to the other or hook up your computer or smartphone to your TV with other kinds of cables to watch everything on the big screen. Now Apple has figured out how to make it easy. It's called iCloud.

What's iCloud?

Everything at Apple starts with an "i." Even Walter Isaacson's best-selling biography of Steve Jobs, called simply *Steve Jobs,* was originally titled *iSteve: The Book of Jobs*. With iCloud, you can automatically sync all your digital books, e-mail, Web bookmarks, movies (purchased from Apple's online iTunes store), music (purchased from Apple's online iTunes store), and recent photos you've taken with your iPad or iPhone so that they all automatically appear on all the other Apple devices you own, without doing anything.

That means that if you own an Apple TV—a device that lets you watch movies and TV shows you've purchased from the Apple store on your TV—you can also automatically transfer your movies, music, and photos from, say, your computer or iPhone to your Apple TV and then play them on your TV.

This magic trick presupposes one thing: that you have a wireless Internet connection in your home. If you do, you can wirelessly stream all this stuff from one Apple product to another. The content is not actually going from one device to another but from a copy stored in the cloud, which is a great big computer storage facility somewhere in the world, to your iPhone or iPad or iWhatever.

There are some nifty things you can do. For instance, you can

- Take a picture with your second- or third-generation iPad[16] and then show the picture to your friends in another city on your iPhone, even if you don't have your iPad with you.

- Listen to music you purchased from Apple on your iPhone at home, then listen to it on your iPad at work, even if you didn't actually transfer the files to your iPhone.

- Take a picture with your iPhone and, a few minutes later, see it magically appear on your computer, where you can touch it up and print it.

- Using iWork, Apple's version of Microsoft Office, create text documents, spreadsheets, or slide shows (generically known as PowerPoints) on your computer, and then automatically see and edit them on your iPad while you're on the train home from work.

- Listen to the songs stored on your computer (or iPhone, iPad, or iPod Touch) you didn't buy from Apple. For $25 per year, Apple will store copies of your own music that it finds in iTunes on its servers, for you to access on your other Apple products.

But I Don't Have iCloud/iPad/iPhone/Mac. What Can I Do?

If you're using a smartphone that's not an Apple product, or a tablet that's not an iPad, you can still sync your documents and photos and music and movies.

Here are two ways to do it:

- **Dropbox: Download the Dropbox app, and then drag files into the special Dropbox folder. Those files automatically sync to the cloud, and from the cloud to every other device on which you've installed Dropbox.**

  Advantages: Free; great for working on a text file at home and then finding the newest version automatically updated on your laptop at work.

  Disadvantages: The actual file (not a copy) resides in your special Dropbox folder, a disadvantage if you like keeping files in specific folders and subfolders. Music files won't automatically appear in an application such as iTunes; you'll have to place them there. Unlike Apple's iCloud, it's not automatic. You actually have to drag a file to a special folder to keep things in sync.

- **SugarSync:** This for-purchase application works with Macs, PCs, Apple products, Androids, BlackBerrys, Windows phones, Symbian phones,[17] and the Kindle Fire tablet.

  Advantage: It works on many different operating systems, so you can sync files from a Windows PC to a Mac, to an Android phone, and so on.

  Disadvantage: You have to pay for it if you want a decent amount of storage. The first 5 gigabytes of storage is free. If you need more, 30 gigabytes will cost you $50 per year.

● ●

Apple's iCloud also gives you 5 gigabytes for free, but Apple doesn't count purchased music, movies, apps, books, TV shows, or your recently taken photos in that limit.

19

Simplify Your Car

Cars used to be as simple to operate as TVs. Put in the key. Start the engine. Go. If your existing car is five years old or more, this may not have hit you yet, but today's vehicles are as connected as the home. E-mail, texting, Blu-ray discs, and Internet access are all available in today's cars. Trying to figure out how to get all that gear fired up, and what you really need, can be as difficult as setting up a wireless modem.

Bluetooth: Necessary or Not?

Bluetooth is a wireless technology that lets you send your cell phone's signal to the car's speaker system. Unlike Wi-Fi, the wireless technology that sends Internet signals around your house, Bluetooth operates only up to 30 feet, so if your cell phone is too far from the car, you won't be able to transfer its signals.

Do I want it? Yes. Bluetooth lets you speak on your cell phone without holding it. Which means you'll be less likely to get yourself killed while you're speaking on the phone because you'll be watching the road and not looking at the phone's screen. The bad news is that many studies show it's just as dangerous to speak on a cell phone whether you're actually holding it or not, so your efforts may come to naught.

With many cars, you can also use Bluetooth to send the music that's stored on your iPhone through the car's speakers. If you have a radio station app, such as the NPR app, you can also listen to distant radio stations through your car radio. It's like having satellite radio for free.

Do I Want (or Need) GPS Navigation?

Yes, with caveats. The biggest benefit is that you'll (almost) never get lost again. With a navigation system you'll know exactly where to turn and where to find the nearest coffee shop, hospital, supermarket, or airport.

You can still get lost even with a GPS system. The instructions about when to turn are not always clear. Do they really mean 50 yards before I turn, or should I make that left sooner? Why does the street onto which I'm supposed to turn have a different name? Why does the main route through the town look like a back alley? This can't be right!

On the other hand, if you love the romance of getting lost, of pulling out a dog-eared paper map and trying to figure out which road you're actually on, of seeing the road you're traveling on in context with your surroundings, that will never happen again once you use a GPS.

Navigation systems are becoming ubiquitous, but getting one that's built into the car can set you back thousands of dollars. There are portable models that can be used in many cars, and they cost just one tenth as much.

Should the GPS Be Built into the Car, or Should I Get a Portable Unit?

There are some advantages to a built-in unit. The screen fits right into the dash, and it incorporates other functions, such as a backup camera or touch controls. Built-in GPS units also have bigger, easier-to-read screens. But the portable, hand-held units are much cheaper, and you can take them with you when you rent a car.

However, the portables are unsightly; with cords dangling, you have to hang one from the windshield and plug it into the cigarette lighter.

GPS on the Cheap

Why buy a GPS unit when you have one already built into your smartphone? GPS apps, both free and low-cost, are available for both Apple and Android phones.

With a smartphone GPS app, you can walk around an unfamiliar city and never get lost. These are the best smartphone GPS apps:

Google Maps: Very good on the iPhone, great on Android. On the iPhone, Google Maps has officially been supplanted by Maps, Apple's own mapping program that, as of this writing, has not proved as useful or as accurate as Google Maps.

MotionX GPS Drive: For the iPhone only, two bucks is all you pay for the basic service, which can help you find your way home or find a coffee shop. You get all the directions you need except voice directions. For $10 per year, you get voice commands and real-time traffic information to help put you on a faster route. Once you get where you're going, you can indicate the spot where you parked in the app, so you can find your car later.

Do I Want a Backup Camera?

Definitely. When you back up, the camera lets you easily see what's behind you. Newer models project lines on a screen to show you exactly when you're about to hit your garbage can (or the small kid playing behind you). With age, it becomes more difficult to turn one's head completely around. Backup cameras do the job for you, and they are due to become standard equipment in all cars in the next few years.

Should I Get a Blind Spot Detection System?

Probably not. As the name suggests, the system beeps when you're inadvertently changing lanes and someone is right in your blind spot. There's a much cheaper solution in wide use throughout the world: convex driver's side mirrors. Just like the "objects are larger than they appear" mirrors on the passenger's side, a convex driver's side mirror widens the angle of view and eliminates the driver's side blind spot.

If you drive a European or Japanese car, you can easily find the correct convex driver's mirror for your car on the Internet. Are you driving American? Manufacturers such as Ford have begun to put a half-convex/half-regular mirror on the driver's side. If you don't have one, buy a small stick-on convex mirror from the shop in the car wash.

I Have Apps on My Smartphone. Why Would I Want Them in My Car?

Just like your smartphone and your TV, cars now come with apps built into the vehicle. For example, with Toyota's enTune system, you can touch the screen to access OpenTable and make restaurant reservations, look up things using the Bing search engine, and listen to music through Pandora, a Web site that lets you create your own personalized channels of music by specifying the type of music or performer you like. Similar services are offered by Audi, Cadillac, Dodge, and others.

If you have apps on your smartphone, you may not need them. You'll probably have to pay a monthly fee for the car versions, but with many contemporary cars, you can access the smartphone apps either by transmitting them to the car with your Bluetooth connection or by connecting the smartphone to the car's entertainment system using the car's built-in port. Depending on the car brand, you may attach your device with a special cable (as found in Audis) or a standard USB cable.

20

Satellite Radio

You'll want this feature, which allows hundreds of specially programmed radio stations to be delivered to your car or, if you have the equipment, to your home, via satellite.

Once there were two separate competing companies called Sirius and XM, but they've now combined and offer the same programming that's hard to beat. Carefully programmed channels offer everything from 1940s music through heavy metal and electronic dance. Old radio shows, sports, cable news, religion, and politics channels are unique. Because it's digital, the sound is crystal clear; delivered via satellite, the music never fades. When you're driving through the Mojave desert, you'll now have more than the two local religious stations from which to choose.

21

HD Radio

Every form of entertainment has become digitized. Movies are now delivered to theaters on hard drives instead of film. Cameras use SD cards, not Kodak film. Digital signals create the super-sharp high-definition images we see on our HDTVs.

Until recently, the only form of entertainment that hadn't become digital was standard old broadcast radio, and now that's changed. Digital broadcast radio—cheekily called HD Radio—delivers the same local channels we've always listened to, but now they're digital.

Just like satellite radio (Sirius/XM), HD radio is static-free; unlike satellite radio, it's still available only in your local area. But digital offers another advantage. The digital signal can actually be compressed—squeezed together to make a smaller signal, so to speak. That means multiple channels can be shoved into the space normally needed, in the analog world, to transmit just one channel. So if you're listening to your favorite local station in HD Radio, you may also be able to get several other stations—called subchannels—that they've tacked onto their main, regular station. For example, a local news station may add several music subchannels to their main offering.

To find out if there are any subchannels, you don't have to do anything. The HD radio will let you tune in to any that are available.

22

Buy This, Not That: The Bottom Line on Consumer Electronics

When buying high-tech electronics, it pays to be choosy.
If you buy everything the salesperson in Best Buy, Sears,
or Walmart recommends, you'll need a second mortgage.
And much of what they recommend you don't need.

One of the great benefits of the digitization of everything
is that the shades of gray that used to define a product have
shrunk. As you may recall from high school, digital is a series
of ones and zeros. Something is either on or off, working or
not, perfect or nothing. That's why an HDTV image doesn't
degrade; it just disappears. Satellite radio signals don't fade;
they shut down in the middle of a song when you go through

a tunnel. On the other hand, because so many digital products are made by a handful of contractors, mostly in China, similar quality is achievable from many different companies.

Gone are the days when a Sony Trinitron represented the pinnacle of television picture quality. Once, Sony sets were discernably different from the competition's. Today, companies such as Panasonic, Sharp, and Sony are getting trounced by startups like Vizio that can purchase the same components that used to be available only to the big boys. So here's the reality: Electronic products continually get less expensive. Similar performance can be found from many manufacturers. But poor build quality still abounds.

Forget the tech specs, the breathless advice from the salespeople, the need to have every single new bell and whistle. Whether it's TV or a digital camera, you don't need to jump for the latest and greatest. Here's a rundown of recommendations that will save you time and money as you make your way through the electronic maze of today's world.

Do I Need an Extended Warranty?

No. As *Consumer Reports* and others (including me) regularly recommend, don't waste your money on extended warranties. Companies such as Best Buy push them hard on consumers for one simple reason: They're big money makers. Given the cutthroat price competition that all electronics retailers face, they need to do whatever they can to eke out every cent possible from their customers.

The truth is that if they're going to fail, most electronic products will do so within the period of the original warranty. There are always exceptions, of course. One model of microwave oven I owned failed twice, each time a year after the factory warranty was up. With the third one I bought the extended warranty; when it failed, I got a free microwave. I bought the same brand repeatedly because it was the only one that fit in the kitchen's space. But one exception does not negate the rule.

How Do I Choose a PC? How Do I Choose a Mac?

If you want to use the Windows operating system, buy a name-brand computer. If you choose a Mac, you can get only a name-brand computer: Apple's.

What Tablet Should I Buy: An Android-Based Product or an Apple iPad?

If you can afford an Apple iPad, that model remains the one to beat. If an iPad is too expensive and you choose one running the Android operating system (made by Samsung, Vizio, and others), make sure the screen is readable even when you tilt it; some cheaper models use LCD screens that must be viewed straight on or the image fades. When you tilt the screen 90 degrees to view it horizontally instead of vertically, the image should alter its position smoothly, not in a delayed, herky-jerky manner.

Does an Android Tablet Offer Enough Apps?

How many is enough? Don't worry about numbers. Think only about the apps you need. It's not practical to list in this book all the best apps because the number and names constantly change. Instead, do a search for "best Android tablet apps" or "best iPad apps" and see whether the ones on offer appeal to you.

How to Choose a Flat-Screen TV

Most of the specs manufacturers use, such as brightness, contrast ratio, screen refresh rate, and megahertz are simply bogus. Unfortunately, there is no standard scale they all subscribe to. Rather, manufacturers have figured out ways to weasel the numbers and make them seem bigger and better than they actually are. So just ignore most numbers. If your TV is 42 inches or larger, make sure it is capable of displaying a resolution of 1,080 pixels (that's one number that can't be made up).

There is no such thing as an LED TV. LED lights are simply used in *LCD* TVs to illuminate the screen. LED *backlit* HDTVs produce a better contrast, and hence a sharper image, than HDTVs that use LED *edge lit* technology. LED backlit HDTVs are also more expensive.

OLED TVs are the next big thing. This technology produces absolutely beautiful pictures on an ultra-thin TV, better and thinner than LCD or plasma sets. The only problem: Right now, OLED TVs are super-expensive. Once OLED TVs drop in price, give them serious consideration.

Digital Cameras

The first HDTV cameras were enormous, expensive pieces of hardware. Today, you can shoot HDTV on an iPhone. The newest smartphones come with enough pixels to create perfectly acceptable, sharp images. And built-in photo manipulation software lets users perform minor edits such as cropping, altering colors, and getting rid of red eye, right in the phone. With a smartphone or many tablets such as the iPad, you can take a picture or an HD-quality video and send it directly to someone's e-mail account or post it on Facebook or Twitter.

With a standard digital camera, you first have to transfer the photos to your computer or tablet and then e-mail or post them. That's why the sale of point-and-shoot digital cameras, the inexpensive models that don't have interchangeable lenses, is plummeting. Smartphones are rapidly becoming the Swiss Army Knife of the digital age. If you're not a professional photographer and just like to take what used to be called snapshots, a modern smartphone is probably all you'll need to do the job.

The Connected Car

Yes, you'll want Bluetooth, but with navigation systems, there are caveats.

Built in navigation systems still cost as much as $2,000. For $200 or less, you can buy a portable navigation system, stick it on your windshield, and take it with you when you travel. Most contain Bluetooth connectivity, giving you Bluetooth features in older vehicles.

Apps? Some of them. Car companies are looking for new ways to get monthly fees from drivers, so they're trying to charge for the use of apps such as Pandora and OpenTable. If you have a smartphone, you can probably transmit the music from its apps to your car's stereo by using Bluetooth. And many portable navigation systems include apps at no extra charge.

Lighting

It's probably not a good idea to switch to compact fluorescents, but you should consider LEDs when replacing lamps that are used frequently. LED lamp prices are dropping precipitously, and they're now available in a wide range of brightnesses and shapes. Soon an LED equivalent will be available for almost every type of lamp.

Phoning to and from Abroad

Here's what you can do to avoid sky-high charges from your cell phone company when you are in another country. Buy a local SIM card if you have an *unlocked* cell phone from AT&T or T-Mobile. Or buy a cheap pay-as-you-go phone from a local cellular company in the country you're visiting; you can typically do this at the arrival airport.

The least expensive way to call other countries from the United States is to use a company such as Pingo or Skype to make calls abroad. Then you can avoid the monthly fees from companies such as AT&T and their higher per-minute fees.

ACKNOWLEDGMENTS

The idea for *Does This Plug into That?* came while I was walking through the Consumer Electronics Show in Las Vegas several years ago. My friend and colleague Danny Abelson noted that many people haven't a clue as to how consumer electronics work; it was his supposition that I would be the perfect person to explain it, for which I thank him.

My agent, Jane Dystel, encouraged the development of the proposal and offered keen insights into the subject material. Chris Schillig, my editor, was very helpful in shaping the work and offering feedback from the perspective of both an accomplished professional and a typical reader. Jay Rubin acted as an important sounding board and reality check.

Throughout the process, my dear wife, Carol, as she has always graciously and lovingly done, supported and encouraged my work. And without the good humor and high energy of my family, the task would have been more arduous. They include my mother-in-law Annabelle, Tanya and Matt, Stacey and Joe, and grandchildren Bella, Hannah, Grace, Ivy, Mia, and Noah.

NOTES

1. If you want an *unbiased* guide to consumer electronics, this is not the book. I have many opinions (e.g., that Apple's products are generally better than the competition's) gleaned over years of writing about technology.
   I express them freely here. You don't have to agree with me.

2. Prometheus stole fire from the gods and gave it to mortals.

3. http://bit.ly/qvzuSk.

4. http://tcrn.ch/NJb40H. **Hint:** Whenever you have a very long Web site link (or URL) that you want to send to someone or quote, shorten it as I did, using the Web site http://www.bit.ly. You cut and paste the original Web site address into the bit.ly page, and bit.ly shortens it for you, for free.

5. http://bit.ly/v9TW1P.

6. To protect your files, server farms are high-security affairs; even their location is often not discussed. The buildings can be huge; Apple's North Carolina facility is said to be 500,000 square feet and to have cost $1 billion.

7. Think it won't happen? Remember how CDs made cassette tapes obsolete? And how 5-inch floppy discs gave way to 3½-inch floppies and then were replaced by CDs? It's only a matter of time before the hard drive disappears as well.

8. They really are the best brownies—simple, and you don't need a mix. is.gd/D4ooDS.

9. One caveat: If you're also using a home theater audio receiver, you may be able to connect all the cables *from* the DVR into the audio receiver and then connect cables *from* the audio receiver

into the TV. By doing this, you can also use the audio receiver as a central control station. For example, you can touch a button on the receiver that might typically be labeled "SAT" to watch satellite TV, or another button labeled "DVD" to watch a disc. Then the audio and video from the appropriate source (DVD player, cable box) will automatically be directed to the HDTV.

10. *Washington Post*; http://wapo.st/PABun1.

11. http://bit.ly/RzmSzW.

12. CDMA stands for Code Division Multiple Access; GSM is Global System for Mobile Communications.

13. SIM stands for Subscriber Identity Module.

14. http://bit.ly/T18Evf.

15. There's an exception to that. You can alter the look of the iPad's screen (as well as other tablets running the Android operating system) in ways that Apple and other manufacturers don't want, if you *jailbreak* your portable device. To do that requires running one of several software applications. Once you jailbreak your iPad, for example, you can install and run applications that Apple hasn't approved. Despite the word used to describe it, jailbreaking of your iPad or iPhone is completely legal
(at least in the United States).

16. The first-generation iPad didn't have a built-in camera.

17. The operating system used by many Nokia smartphones.

INDEX

Index

Index